CAMBRIDGE LIBRARY COLLECTION
Books of enduring scholarly value

Technology

The focus of this series is engineering, broadly construed. It covers technological innovation from a range of periods and cultures, but centres on the technological achievements of the industrial era in the West, particularly in the nineteenth century, as understood by their contemporaries. Infrastructure is one major focus, covering the building of railways and canals, bridges and tunnels, land drainage, the laying of submarine cables, and the construction of docks and lighthouses. Other key topics include developments in industrial and manufacturing fields such as mining technology, the production of iron and steel, the use of steam power, and chemical processes such as photography and textile dyes.

The Drainage of Fens and Low Lands

A respected hydraulic engineer, William Henry Wheeler (1832–1915) established himself as an authority in the fields of low-lying land reclamation and drainage, embanking, and the improvement of tidal rivers and harbours. Based on articles written for *The Engineer* in 1887, he published this more substantial work in 1888 to provide a practical point of reference for those working on existing drainage systems or designing and implementing new schemes. Drawing on first-hand knowledge of the low-lying fens of England and the polders of Holland, Wheeler describes the processes of drainage by gravitation and steam power. The book also includes chapters on lifting and draining water using the scoop wheel, the Archimedean screw pump and the centrifugal pump. Providing also a wealth of operational facts relating to pumping stations in England and abroad, this remains a rich source of information for those interested in the history of civil engineering.

Cambridge University Press has long been a pioneer in the reissuing of out-of-print titles from its own backlist, producing digital reprints of books that are still sought after by scholars and students but could not be reprinted economically using traditional technology. The Cambridge Library Collection extends this activity to a wider range of books which are still of importance to researchers and professionals, either for the source material they contain, or as landmarks in the history of their academic discipline.

Drawing from the world-renowned collections in the Cambridge University Library and other partner libraries, and guided by the advice of experts in each subject area, Cambridge University Press is using state-of-the-art scanning machines in its own Printing House to capture the content of each book selected for inclusion. The files are processed to give a consistently clear, crisp image, and the books finished to the high quality standard for which the Press is recognised around the world. The latest print-on-demand technology ensures that the books will remain available indefinitely, and that orders for single or multiple copies can quickly be supplied.

The Cambridge Library Collection brings back to life books of enduring scholarly value (including out-of-copyright works originally issued by other publishers) across a wide range of disciplines in the humanities and social sciences and in science and technology.

The Drainage of Fens and Low Lands

By Gravitation and Steam Power

WILLIAM HENRY WHEELER

CAMBRIDGE
UNIVERSITY PRESS

University Printing House, Cambridge, CB2 8BS, United Kingdom

Published in the United States of America by Cambridge University Press, New York

Cambridge University Press is part of the University of Cambridge.
It furthers the University's mission by disseminating knowledge in the pursuit of
education, learning and research at the highest international levels of excellence.

www.cambridge.org
Information on this title: www.cambridge.org/9781108066402

© in this compilation Cambridge University Press 2013

This edition first published 1888
This digitally printed version 2013

ISBN 978-1-108-06640-2 Paperback

This book reproduces the text of the original edition. The content and language reflect
the beliefs, practices and terminology of their time, and have not been updated.

Cambridge University Press wishes to make clear that the book, unless originally published
by Cambridge, is not being republished by, in association or collaboration with, or
with the endorsement or approval of, the original publisher or its successors in title.

THE DRAINAGE

OF

FENS AND LOW LANDS

BY GRAVITATION AND STEAM POWER.

BY

W. H. WHEELER, M. Inst. C.E.,

BOSTON, LINCOLNSHIRE.

E. & F. N. SPON, 125, STRAND, LONDON.

NEW YORK: 12, CORTLANDT STREET.

1888.

PREFACE.

THE drainage of fens, polders, and low lands, although a matter of great importance, is a subject on which it is difficult to obtain information. There is no complete treatise on the drainage of low land by gravitation or by steam-power to which an engineer called in to advise can refer. Numerous papers describing drainage and reclamation works are to be found in the 'Transactions of the Institution of Civil Engineers' and in the engineering journals, yet these are too scattered to be of ready service; and there are many important drainage works of which no record exists. The Author has endeavoured in the following pages to supply this want.

The lands dealt with are such as lie below the level of ordinary high water of the sea, and therefore require embanking, and special contrivances for securing proper outfalls for their drainage. These lands may be divided into four classes. First, those that are situated at a sufficiently high level as to be able to discharge their drainage during low water; these require only protecting from inundation at high water by embankments, and the main drains by sluices, which will discharge the contents of the drains at low water, but close against the rising tide. Second, lands at a lower level, which can obtain a discharge by gravitation under ordinary circumstances, but require that the water should be lifted out of the main drain by mechanical power during floods. Third, lands lying beneath low-water level, or so low that mechanical agency is required during that portion of the year when the rainfall exceeds the evaporation. Fourth, land, adjacent to the upper reaches of tidal rivers, the surface of which is sufficiently high to obtain efficient drainage by gravitation, but owing to the distance from the sea, the cost of deepening and improving the channel of the river, or the length of main drain required to be cut to carry the water to a lower outfall, frequently through high intervening land, makes the cost of drainage by gravitation greater than that by steam-power.

The greater part of the matter contained in this book, and most of the illustrations, appeared during the year 1887 as a series of articles in *The Engineer*, entitled 'The Drainage of Fens and Low Lands by Steam Power,' and the Author here acknowledges the kindness of the Editor of that journal in allowing him the use of the engravings.

The information given has been obtained from various sources—partly from descriptions of drainage works contained in engineering publications, and from well-known works on hydrology. A great portion of the matter is entirely new. The Author's own experience as an engineer living in a fen country combined with a personal inspection of all the principal drainage works in England and Holland, has given him the opportunity of collecting together a large number of facts, a record of which he hopes will be of service to those having to design new works, or the superintendence of those already in existence. For the ready assistance of those who have aided him in the matter, the Author here tenders his best thanks.

In order to facilitate comparison of the different kinds of machinery in use for lifting water, the results are in every case reduced to one common standard in English measures of tons of water lifted to a given height in feet, and of work done in water lifted and discharged, expressed by the term water-horse-power (W.H.P.).

In the Appendix will be found several useful formulæ for making hydraulic calculations. As the works in other countries are almost invariably given in French weights and measures, the equivalents of these in English terms are also given in one of the tables.

CONTENTS.

CHAP.		PAGE
	PREFACE..	iii
I.	INTRODUCTION	1
II.	DRAINAGE BY GRAVITATION	7
III.	FIELD DRAINAGE	36
IV.	DRAINAGE BY STEAM POWER..	50
V.	THE SCOOP WHEEL	70
VI.	THE ARCHIMEDEAN SCREW PUMP..	88
VII.	THE CENTRIFUGAL PUMP..	92
VIII.	DESCRIPTION OF PUMPING STATIONS..	106

APPENDIX.

TABLE		
I.	—Explanation of terms used..	158
II.	—Weights and measures relating to water; French weights and measures, and their equivalents in English terms	158
III.	—Rules for ascertaining the pressure and velocity of water	160
IV.	—Formula for finding the velocity of water in open channels	161
V.	—For finding the quantity of water due to rainfall discharged off the land, and the horse-power required for raising the same	162
VI.	—For finding the discharge of water through sluices	162
VII.	—For finding the quantity of water due to twenty-four hours' rainfall if discharged in a limited time owing to a sluice being closed by the tide	163
VIII.	—For ascertaining the time the doors of a sluice are closed by the tide	164
IX.	—For ascertaining the height of the tide at any hour during ebb and flood	165
X.	—For finding the horse-power required for pumping engines	166
XI.	—Showing the dimensions and capacity of drains and sluices, and the area of land for which they are adapted	167

LIST OF ILLUSTRATIONS.

	PLATE
Sea Sluice for Drain, width 8 feet opening	1
Ditto, with three openings	2
Diagram showing Angles of Ingress and Egress of Scoops	2
Centrifugal Pump and Engine at Messingham	2
Scoop Wheel, Plan and Section	3
Archimedean Pump	3
Sections showing Blades of Pumps..	4
Centrifugal Pump and Engine at Grootslag Polder	4
Ditto, Legmeer Polder	4
Scoop Wheel at Podehole..	5
Centrifugal Pumps at Lade Bank	5
Diagram of Engine at 100 foot..	6
Whittlesea Mere Pump Station	6
Section of Pump at Burnt Fen..	6
Scoop Wheel at Nordelph..	6
Plan and Elevation of Pumping Station at Ferrara	7
Pump at Minden in Holland	7
Plan and Sectional Elevation of Pumps at Lake Haarlem..	8

THE DRAINAGE
OF
FENS AND LOW LANDS.

CHAPTER I.

INTRODUCTION.

THE object of this book is to give a general description of the works and machines used in draining low land, with such information and practical hints as may be of service to those having to superintend drainage districts or to design new works. Living in the midst of the Fenland, where, with the exception of Holland, the science of drainage has been applied to the improvement of land on a larger scale than any other part of the world, and having been engaged professionally as an hydraulic engineer for the last twenty-five years, the author has had the opportunity of collecting together a large amount of facts and experience relating to this particular subject, and hopes that, by rendering these accessible to others who have not had the same opportunities, he may be doing some service.

A civil engineer called in to advise as to drainage has not, as a rule, had the experience necessary to enable him to design the details of machinery required for the purpose, but he ought to have sufficient knowledge of the subject to enable him to judge as to what type of machinery is best suited for the particular work required, and to be able to draw up such a specification that tenders for the work may be furnished upon a similar basis. While a certain amount of freedom as to details may be allowed to the makers tendering for the machinery, the responsibility of the successful carrying out of the work must rest with the engineer.

The want of sufficient knowledge of this special subject, and experience of what has been done, has led in several cases to the erection of unsuitable machinery, waste of money, and failure to effect the object required in the most effective and economic manner.

Very large tracts of rich land suitable for cultivation lie at so slight an elevation above the sea, that this land can only be rendered fit for cultivation by artificial means. The Polders in Holland, the Fens and marsh land in England, the sites of old lakes in Italy and other parts of Europe, and considerable quantities of low land in the delta of the Mississippi and in South America and the colonies, have been brought into cultivation and made to yield large quantities of produce by a complete system of drainage. In India, Egypt, and California, on the other hand, large areas of land are rendered fertile, and made to grow abundant crops of fruit and other produce, by means of irrigation; for this purpose, besides channels carried from the rivers in the higher districts, mechanical agency is largely resorted to for lifting the water into the irrigating channels. The steam engine and centrifugal pump are now rapidly replacing the more ancient and simple means of lifting the water.

In Holland the reclaimed land, consisting, for the most part, of the beds of lakes, is at so low a level that pumping has been almost universally resorted to.

In the valley of the Po upwards of 600,000 acres of marshy land have been drained and transformed into rich country by means of pumps and lifting wheels, the drainage of which was only practicable since the introduction of steam power, owing to the difficulty of getting rid of the water by gravitation. In the south of France also large tracts of land have been made available for cultivation by reclamation and steam drainage.

Primarily drainage by gravitation is the simplest and most effective method. When windmills were the only motive power, this means of getting rid of the water was more of a necessity than it is at the present time, when pumping engines can be obtained with a high degree of efficiency, and can be worked at a small cost.

The engineers who originally designed the drainage for the Fenland in England endeavoured by means of long straight cuts to obtain a natural outfall into the sea, or main rivers, at low water, excluding the tidal flow by sluices having self-acting doors. Modern engineers have followed the example thus set, and consequently, while large sums have been spent in artificial cuts with the object of obtaining drainage without the aid of mechanical power, the improvement of the main tidal outfall rivers has been more or less neglected, and the chimneys of pumping engines are everywhere to be seen scattered over the fens.

In no instance in the Fenland has the attempt to obtain what is called a "natural drainage," that is, drainage by gravitation, been completely successful. While the higher land is well drained, the lower fens, which often lie at the greatest distance from the outfall, can only be kept fit for cultivation by lifting the water out of their drains. The lift of the water, and consequently the cost of the pumping, has been considerably reduced, but the taxes to meet the interest on the outlay for the works, in addition to the cost of pumping, is much higher than in adjacent districts where more reliance has been placed on pumping. Fen land, which was well drained when the main outfall drains were first constructed, afterwards had to resort to pumping, owing to the lowering of the surface from the consolidation and shrinking of the peat. Such has been the case in the East Fen in

Lincolnshire, where a district of 30,000 acres, formerly drained by gravitation, is now kept free from water by engines driving centrifugal pumps erected in 1868, a description of which will be given further on.

In the Black Sluice district the main drain, twenty-one miles in length, was enlarged and deepened, and a complete system of internal drainage carried out in 1848, with the hope that the fen land would by this means be effectually drained. One district after another has, however, resorted to pumping as the only means of giving complete relief to flood water, till now all the low land at the upper end of the main drain is kept free from water in floods by steam-power. On the river Witham, fourteen districts, containing an area of over 30,000 acres, are drained by steam-power.* In the North Level of the Bedford Level, where the taxation is already very high, owing to the large amount expended in erecting outfall sluices and perfecting the main drains, the lower districts suffer greatly from flooding in wet seasons, and power was obtained in the session of 1881 to erect a large pumping station for the better drainage of the district.

An enormous outlay was incurred in the Middle Level in Cambridgeshire in cutting a large main drain eleven miles long with an outfall sluice discharging into the river Ouse at a point nine miles lower down the channel than where the old drain discharged. It was considered at the time that this drain would afford such a good discharge for the water that pumping would become unnecessary. The numerous engines which are now at work in this level prove that these expectations were not justified by the result. Although the cost of lifting the water has been reduced, yet nine-tenths of the land has to be secured from flooding in wet seasons by mechanical means. There are certain districts in this level which

* 'The Fens of South Lincolnshire,' by W. H. Wheeler, C.E. Simpkin and Marshall. This book is now out of print. A new edition is in course of preparation, and will shortly be issued by the same publishers.

refused at the time to be brought into the new system, preferring to rely on their engines and pumps. The taxation for paying for the maintenance of the pumping stations in these districts is so light, compared with those which are in the rest of the Middle Level which have to contribute to the outfall-tax, as fully to have justified the opposition to the scheme, and proves conclusively that there are instances where drainage by gravitation is often more expensive than drainage by steam power.

In the South Level, owing to the want of improvement in the river Ouse, the main outfall of the district, the discharge from the main drains is so defective in wet seasons as to prevent the floods getting away with sufficient rapidity. The water consequently rises to an undue height in the river above the tidal outfall at Denver, bringing pressure on the banks beyond what they were intended to stand, a greater head to pump against, and additional water from leakage to be lifted. The question as to whether the better course for the improvement of the drainage of this level would be to imitate the example of the Middle Level, and make a new cut discharging nine miles lower down the river than the present outfall sluice at Denver, or to improve the outfall and its discharging capacity, and so lower the height of the water in floods within safe limits, and still rely on steam drainage, was referred to the author to report on. The conclusion arrived at, after a thorough investigation of the subject, was that efficient drainage of the low fen lands in this district by gravitation was not advisable, and that the cost of draining by steam power would be less than by gravitation.* The interest on the outlay for the money required for carrying out the gravitation scheme would have put a greater tax by about two shillings an acre on the land than that required for the improve-

* 'Report on the Improvement of the River Ouse, between Denver Sluice and the Eau Brink Cut.' By W. H. Wheeler, M. Inst. C.E. February 1884. Library, Inst. C.E.

ment of the present outfall and the continuance of steam power.

The great improvements which have been made in the steam engine and water-raising machines, together with the greater facilities for obtaining and the lower price of coals, have very considerably reduced the cost of lifting water as compared to what it was when many of the improvements for the drainage of the fen land were carried out. There is no doubt if the work had to be done now, the engineers engaged in those works would have trusted more to mechanical lifting than to gravitation.

The choice as between gravitation and steam power for draining low lands resolves itself into a question of cost. If the annual charge for interest on the outlay for a gravitation scheme, with a proportionate sum for repayment of the principal, exceeds the average annual cost for a pumping station, then the steam power is decidedly preferable, not only as being more economical, but as rendering the district more thoroughly independent of outside circumstances. The annual charge for a gravitation scheme is constant, be the season wet or dry; whereas a pumping station adapts itself more readily to the actual work to be done, the charge for coals varying with the amount of water to be pumped.

One obstacle to the more general use of steam power has been the excessive cost of pumping stations in some localities from the use of imperfect machinery, and ignorance on the part of those concerned in the management. While some engines and pumps are so efficiently designed and managed, as to leave little or no room for improvement, others are being run with a most extravagant use of coals, and imposing a rate of taxation for their maintenance that is quite uncalled for.

CHAPTER II.

DRAINAGE BY GRAVITATION.

THE drainage of every district will vary either with the elevation of the ground, the nature of the soil, the distance of the district from the outfall, and other circumstances. It is frequently the case that a low fen or polder has not only the rainfall due to its own area to contend with, but also that of the adjacent higher land. When the water has to be pumped this is a serious consideration; and even when it flows away by gravitation, the water coming from the higher land will over-ride the low-land drainage and frequently be the cause of flooding or of obstruction to the free discharge of the low-land water. To obviate this, whenever practicable, a drain surrounding the district is made, which collects the high-land water and prevents it finding its way into the lower level. These drains were termed by Mr. Rennie "catch-water drains"; and the same purpose is attained in Holland by a dyke and canal termed the "ringvart."

RAINFALL.—If the district to be dealt with consists purely of low land, provision must be made for the whole of the rain which falls on the surface in winter, as from the low level of the land none of the rain can soak away. In winter the loss by evaporation and absorption by vegetation in wet weather amounts to scarcely any perceptible quantity, and with the outfall stream probably bank-full, and all the surrounding land saturated, no deduction can be made for natural soakage. If the district consists partly of high land, the proportionate area of the drains can be made less, as some of the rain of the higher districts will be absorbed by the soil on which it falls, the quantity depending on its nature. If the higher lands consist of either sandstone or limestone, or especially if of chalk, the absorption, after heavy rain, will be very great, and the discharge off the surface to the drain very small. In some chalk districts the absorption is so great that there is an entire absence of streams; in others, the discharge is so small, that the waterway of the bridges and culverts is not an eighth of that required in clay districts.

A large fall of rain at any particular period does not necessarily produce a flood in flat districts. When rainfall succeeds a season of dry weather, it takes some time to saturate the land; and owing to the large capacity of the arterial drains, their small declivity, and the level character of the land, it occupies some time before these become fully charged. On the other hand, if rain falls after a continuance of wet weather, the land, together with the drains, is already charged, and any exceptionally heavy rain, although even for a short duration, is at once succeeded by a flood, as the drains cannot carry off the excess in their surcharged state.

In the fen districts of Lincolnshire the average rainfall of recent wet years has been 32·39 inches, of which 17·52 inches was due to the six winter months, September to February, which, spread over this period, gives an average daily rainfall of 0·097. Taking the periods of excessive rain which occurred during the same time, extending over six to thirty successive days, the greatest average fall per day has been 0·41 inch for fourteen days in October 1883 and November 1885, the next highest being 0·29 for six days in February 1883. The average mean rainfall during the twenty-one floods since 1852 was 0·26 for seventeen days.

The quantity of drainage which was provided for by the old Fen engineers was that due to the water arising from a continuous rainfall of a quarter of an inch of rain in twenty-four hours, making no deductions for soakage or evaporation. This calculation was also adopted by Sir John Hawkshaw

Drainage by Gravitation.

for the engines erected for the drainage of the East Fen in Lincolnshire, and by Sir John Coode as the maximum quantity to be lifted by the engines proposed to be erected for the drainage of the North Level.

Taking the rainfall in the Fen district as a guide, it may be estimated that provision should be made for a daily rainfall equal to about ·0076 of the average annual rainfall of wet seasons.

In Holland, the quantity generally calculated as having to be lifted off the Polders varies from a quarter to three-eighths of an inch of rain in twenty-four hours. The rainfall at Lake Haarlem averaged 31·27 inches for the ten years ending 1872, and 32 inches for the ten years ending 1886, the greatest fall being 39·13 inches in 1877, the engines running 8056 hours that year, against 6823, the average for the ten years.

In Ireland, for the Rathdowney drainage on the river Erkina, provision was made in the channel for a discharge of 600 cubic feet per square mile per minute, equal to a continuous rainfall of about three-eighths of an inch in twenty-four hours. The soil was a deep alluvium, the drainage area containing 56·7 square miles of low-lying land on the lower limestone.* At the Wexford Harbour Reclamation Works, where the rainfall of wet years amounts to over 50 inches, and the mean from 45 to 48 inches, it was estimated that three-fourths of this quantity, or 34·2 inches, would have to be pumped; but the machinery was made of sufficient capacity to lift nearly an inch of rainfall in twenty-four hours.

TIDAL OUTFALLS.—When the outfall of a main drain is into a tidal stream, it has not only to be capable of discharging all the water due to the rainfall, but also must be able to discharge this water within a limited time, the doors of the sluice being closed by the tide for a certain period twice a day. This element has to be taken into consideration in

* Trans. Inst. C.E., vol. lix. p. 265.

determining the sectional area. The time the sluice is closed by the tide should be ascertained by observation, but it may be found approximately from Table VIII. in the Appendix.

DEPTH AND WIDTH OF DRAINS.—The depth of a drain must be determined principally by the level of the land with reference to low water in the outfall in floods, and by the depth it is thought desirable to maintain the water in the ditches below the surface of the land. This should not be less, where obtainable, than 3 feet or 3 feet 6 inches, to allow the under drains to run clear. In peat land, two feet below the surface is considered sufficient for the water-level. Taking the depth to be 3 feet 6 inches, and allowing for one foot of water in field ditches, the bottom of these will require to be 4 feet 6 inches below the surface. The depth of the main drains and level of the water must be such as to provide for the drainage of the lowest land situated the greatest distance from the outfall, and also to allow for the necessary fall in the surface of the water from this land to the outfall. The mean width of the drains must be determined by dividing the area required by the attainable depth.

The greater the depth within certain limits in proportion to the width, the better will be the discharging power of the drain. The greater the body of water as compared to the area of rubbing surface, the more free is the water to move. The proportion of rubbing surface of the sides and bottom to the area of the water is termed the hydraulic mean depth, and is found by dividing the latter by the length of the former, or the area by the wetted perimeter.

The best form of channel for conveying water is when its hydraulic mean depth is at a maximum, and this is attained when the mean width is double the depth, or a semicircle with the diameter for the water-line. Such a form, however, is never attained in land drains, and the width will generally be from four to six times the depth.

MOTION OF WATER.—The motive power which causes

water to flow to its ultimate destination, the sea, is that due to gravity, each particle of the fluid endeavouring to attain the lowest level. If the bottom of the drain be perfectly flat, so long as there is any fall in the surface of the water the particles at the higher end will continue to move until they have arrived at the lower level; and as every particle is free to move, the whole mass of water in a channel, from the surface to the bottom, will be in motion until a low-level horizontal surface is attained. The action of a mass of water in a running stream is not that of a body moving in a plane parallel with the surface or bottom of the stream, but partakes more of a rotary character, by which the particles of the water are continually being rolled round and round from the bottom to the top, this action being increased where the bottom is irregular and full of holes. The result of this rotary motion is indicated by the numerous miniature whirlpools that are constantly forming on the surface, and by the manner in which light floating substances appear and disappear. It is, no doubt, due to this centrifugal motion that the deep holes that are found in running streams are maintained, the pebbles being whirled round and round at the bottom, keeping the loose soil continually in motion and allowing it to be carried away by the running water.

A very slight inclination in the surface of water, even the fraction of an inch in a mile, will cause movement. An increase in the surface inclination rapidly increases the velocity and the quantity discharged. An increase of four times the fall in the surface of a stream doubles its discharge.

VELOCITY OF WATER IN DRAINS.—The velocity is governed by the rate of inclination of the surface of the water, and not of that at the bottom of the drain. The velocity of the water due to gravity is checked and retarded by the friction of the water against the rubbing surfaces with which it comes in contact, this rubbing surface consisting of the sides and bottom of the drain, weeds, sides of bridges, or other impediments. A deep stream, therefore, having less rubbing surface, has a greater velocity for the same inclination than a shallow one, and a drain with a regular channel will discharge more water than one with shoals and depressions in its bottom, or having frequent bends. The velocity of a stream is in proportion to the square root of the depth. Inclination combined with depth constitutes the power required to overcome obstacles, and a deep stream, from its greater gravity, will have more scouring effect than a shallow one having the same velocity. Every particle of water in a flowing stream being free to move, the whole body of water from the surface to the bottom is in motion; but owing to the retarding influence of the bottom and sides, the velocity is greatest in the centre of the stream, and at a small distance below the surface, and is least along the bottom of the channel. The mean velocity of a stream is generally taken as about four-fifths of the surface velocity in the centre. If the same quantity of water that enters a stream leaves it at the lower end, it is evident that the same body of water must pass throughout its whole length, whatever the difference in the area of the section at different parts; the water increasing in velocity when the area is small, and decreasing when it is large, the surface inclination varying in proportion.

The object to be sought in laying out a drain in a flat country is to provide a channel wherein the water shall be moved along its intended course with such ease that as small an inclination and area shall be used as possible. Every increase beyond what is absolutely necessary is a waste of land and expense in excavation.

The method of calculating the velocity of a stream is by a formula, deduced from the effect due to the action of gravity, reduced by the amount of friction encountered. The theoretical velocity V, is found by multiplying the square root of the product of the hydraulic mean depth R, by twice the fall in feet per mile. The result must be reduced by

Drainage by Gravitation.

a coefficient varying with the nature of the stream, and determined originally by experiment checked by practice. $V = (\sqrt{R \times 2F})$ C. V equals mean velocity in feet per second; R equals hydraulic mean depth in feet; F, fall of surface in one mile in feet; C, a constant, varying from 0·90 in rivers and large streams with considerable depth of water to 0·60 for small drains in good order. The surface velocity in the centre of the stream, as found by floats or by a current-meter, must be multiplied by 0·83 to find the mean velocity of ordinary drains.

The amount of retardation caused by friction, is an element that can only be determined by experiment. Authorities place this figure as low as 4 per cent. for streams discharging a large volume of water, say from 2000 to 3000 cubic feet a second, and as high as 40 per cent. for small streams of one-hundredth this capacity. From observations deduced by experience and comparison of the various formulæ given by Beardmore, Stevenson, Neville, and others, the following may be taken as the factors by which the theoretical discharge is to be multiplied to allow for friction, and as generally applicable to drains in low flat districts.

	Depth. feet.	Mean Width. feet.	Factor.
Main drains	6	30	·80
Secondary drains	4	20	·70
Small drains	2	10	·60

WEEDS.—If weeds are allowed to grow in the drains considerable deduction must be made from the theoretical calculation, not only from the fact that they abstract from the area of the drain, but necessarily retard the flow of the water. The actual amount of deduction will vary with the quantity and nature of the weeds, but a rule is given by Neville which provides that the hydraulic mean depth of the drain should be multiplied by 1·70.

The following is an illustration of the way in which weeds hold up the water. In the river Hull in Yorkshire, over a distance of 5¾ miles between Hempholme Lock and the

South Bullock Pumping Station, which is about 12 miles above the junction of this river with the Humber, in July 1887, previous to the cutting of the weeds, the mean inclination in the surface at low water was at the rate of 0·63 feet per mile. After the weeds were cut between these stations and in the river below, this was reduced to 0·15 feet per mile, the water standing at Hempholme Lock, at low water, 2·84 feet lower after all the weeds were cut than it did before the cutting commenced. The lowering of the level and the decrease in the rate of inclination was progressive as the cutting went on. The difference was not so great at the lower station, as the weeds below this were not so thick, and the river more affected by tidal influence; but there the low-water level was depressed 1·09 feet by the removal of the weeds. This alteration was considered to be due almost entirely to the removal of the weeds.

INCLINATION IN THE SURFACE OF WATER.—Every inch of fall being of value in flat districts the area of the drain is so proportioned that the surface inclination of the water in the drain is reduced to as low a rate as is compatible to the efficient discharge of the water. Dubuat considered that the eighth of an inch in a mile would cause a sensible movement in a canal. The inclination of main tidal rivers through flat districts, when properly trained, varies from 3 to 12 inches in a mile, and should not under ordinary conditions exceed 6 inches. Large tidal channels, when undisturbed by floods, flow with an inclination as small as 1 inch in a mile. Main arterial drains in well-drained flat districts flow with an inclination of from $1\frac{1}{2}$ to 3 inches per mile. The Middle Level drain, when the syphons were at work, had a surface inclination, over 15 miles at the lower end, of $1\frac{1}{2}$ inch per mile. The main drain in Deeping Fen, Lincolnshire, discharges the full quantity required to keep both the pumping engines at work, with an inclination of 8 inches in 12 miles, or at the rate of $1\frac{1}{2}$ inch per mile. The Black Sluice drain in the same county, in floods, has an inclination of

Drainage by Gravitation.

3¼ inches per mile for the first 12 miles above the outfall. The surface inclination of the canalised part of the river Witham is about 3½ inches per mile. The tidal portion of the river Ouse between the upper end of the Eau Brink Cut and Denver Sluice, which has a very irregular section, has a mean fall in floods of 11 inches per mile, the greatest fall being at the rate of 17 inches for three-quarters of a mile round a sharp bend in the river, falling to 3 inches per mile through the Eau Brink Cut to Lynn. The inclination of the Shannon, where it passes through flat land, is at the rate of 2½ inches per mile. In the main drains and canals in the polders in Holland the surface inclination varies from 1½ to 3 inches in a mile.

SLOPE OF SIDE OF DRAIN.—The determination of the slope, or batter, to be given to the sides must depend on the nature of the soil, but will also be guided to a certain extent by the width and depth of the drain. It is frequently advantageous to lay out the slopes of the main outfall drains at a flatter batter than the nature of the soil absolutely requires, in order to afford larger storage room during the time the flow of the water is stopped by the tide, the rapid increase in the area above the mean water line affording a large reservoir for the water, without increasing the width at the bottom. On the other hand, to make the slopes more than necessary entails useless expense. Soils of a sandy nature will require very flat slopes, as the soil is easily removed by the wash of the water. In the alluvial soils of marsh districts the large drains may frequently be found having batters of only one to one. Clay, according to its nature, will stand with a small slope, so far as the wash is concerned, but is liable to slip if laid too steep. Peat requires very little batter, in fact, many of the secondary drains in the peat districts may be seen with their sides almost vertical. A careful observation of the existing drains or watercourses in the neighbourhood, and the slope to which they have adapted themselves, will form a guide as to the section to be given to a new drain.

Main drains seldom have less slopes than 2 to 1, and this

is increased to 3 to 1 in light soils, easily washed away, or where the substratum is soft and will not bear the weight of the sides after the drain is excavated.

In excavating drains where the substratum consists of soft alluvial soil, frequently overlain by material of a denser and more stable character, the bottom will spring up and let the sides down. When ground of this character has to be encountered, the soil excavated must be moved to a sufficient distance to prevent its weight forcing the sides into the cutting, and should not be placed nearer than six feet. It may even be necessary in very bad places to build the sides up with fascines to prevent them slipping in.

As the drains diminish in size the slopes may decrease to $1\frac{1}{2}$ to $1\frac{1}{4}$ to 1 for the second class drains, down to $\frac{3}{4}$ to 1 for the main ditches.

AREA OF LAND OCCUPIED BY DRAINS.—The area of land occupied by drains varies considerably with the position and character of the land and its means of discharge. In the low polders in Holland, drained by windmills, it is estimated that one-tenth of the land is occupied by water. If drained by steam power, one-twentieth is frequently occupied. The following proportions of the area of drains to the total area of land of the principal polders in Holland is given in Huet's 'Stoombemaling van Polders en Boezems':—

Polders.	Area, acres.	Proportion.
De Beemster	18,824	$\frac{1}{13\cdot4}$
Haarlem Meer	44,700	$\frac{1}{20}$
Rijnland	305,000	$\frac{1}{27}$
Kennemerland	16,000	$\frac{1}{38}$
Gedeete van Vijf	18,000	$\frac{1}{42\cdot5}$
Amstelland	74,100	$\frac{1}{59}$
Defland	74,100	$\frac{1}{77}$
Waterschap de Rotte in Schieland	27,200	$\frac{1}{85}$ *

* 'Stoombemaling van Polders en Boezems,' door A. Huet, C.E. 's Gravenhage, 1885.

When land drained by gravitation discharges its water into a tidal stream, or, if drained by steam, the pumps do not work at night, the area of the drains will require to be larger than where the discharge is constant, as a reservoir has to be provided for the accumulation of water during the time the doors are closed by the tide, or the pumps are not at work.

CLEANING DRAINS AND REMOVAL OF WEEDS.—Drains running through fens and flat districts, where the current is never very rapid, and where generally in summer-time there is no current at all, are liable to become choked with weeds. The earthy matter carried by the water in suspension in floods is arrested by these weeds, and gradually a deposit accumulates at the bottom of the drains, in which more weeds grow, and so accretion goes on. The uniform depth of the channel is thus deranged, the bed of the river rises and consequently the water-way and the discharging capacity of the drain is diminished. The weeds themselves also prove a great obstruction to the flow of the water. Accumulation of deposit also takes place across the main drains at the places where the lateral drains come into them.

It is generally the practice to cut the weeds twice or three times a year. The ordinary method is by an implement resembling a number of scythe blades joined together, which is drawn backwards and forwards across the drain by men stationed on either side, and working upwards against the stream. The weeds as cut are drawn out by rakes, and placed above the highest flood level. The cost of this work in the fen district, where it is termed "roding," is about 20*s*. a mile for drains from 12 to 20 feet wide, and 30*s*. for larger drains.

A more effectual plan, and one which also at the same time removes shoals and accumulations of deposit, is by loosening and breaking up the deposit by means of a revolving implement drawn along the bottom of the drain at the time when a current is running down. By this means not only the soil of the shoals is loosened and carried away in suspension, but also the roots of the weeds are torn up and are carried by the current out of the drain. If this is done frequently, drains can be successfully kept clear of weeds and deposit; and may even be deepened at considerably less cost than by dredging, or by spade labour, without any injury to the outfall. Too little advantage is taken of the capacity of the water as a carrying agent in the improvement of rivers. In dredging, the chief expense and difficulty is the removal of the material dredged up. By continually stirring up the matter to be removed, it rises in the form of mud, and the particles are sufficiently small to be moved and carried away in suspension. If the section of the channel is uniform, the velocity of the water will carry the material entirely away, but wherever there are wide places and slack currents, there will be a tendency for the matter in suspension to be deposited. By frequently and continually running the machine up and down the drain within the defined limits of the waterway a uniform and regular channel can be maintained free from shoals and weeds.

TRANSPORTING POWER OF WATER.—The transporting power of water may be realised by considering the turbid condition and immense quantity of matter carried down by comparatively sluggish streams in times of flood. The quantity of material transported by such rivers as the Humber and the Trent is evidenced by the fact that the warping lands on to which the water is allowed to flow are raised by the alluvium which subsides, as much as two feet in one year.

Owing to the constant change in the direction or motion of the water causing horizontal and vertical eddies there is a considerable upward vertical action, which counteracts the downward motion of particles of matter of heavier specific gravity carried in suspension. Thus particles of soil are kept suspended, which in still water would fall to the bottom. In addition to the matter carried in suspension, the action of the

Drainage by Gravitation.

water rolls along the bed of the channel, particles of material the specific gravity of which is too great to allow of their being raised above the bed of the channel. A stream running with a velocity of 6 inches a second, or about one-third of a mile an hour, will transport soft clay; a velocity of half a mile an hour will carry sand as large as linseed; a velocity of two-thirds of a mile will sweep along fine gravel; while a current moving at the rate of a mile and a half an hour will roll along rounded pebbles; and at the rate of two miles an hour pebbles the size of a hen's egg will be moved along the bottom of the channel.

In some rivers upwards of 2 per cent. in weight of the total volume of water passing along their channels consists of material carried in suspension. In the Tees, when the draining works were going on, the quantity of material in suspension in the water was as much as 5 oz. to a gallon, or $\frac{1}{32}$ of the weight of the water. The proportion in the Durance and the Vistula in floods is $\frac{1}{48}$; in the Garonne and the Rhine in Holland, $\frac{1}{100}$; the Rhone, $\frac{1}{230}$; the Po, $\frac{1}{300}$; in the Humber, $\frac{1}{219}$; in other rivers the proportion varies from the above as a maximum to $\frac{1}{17000}$, as a dry weather flow. To give an illustration of the quantity of material transported by a river, it is stated that the Durance transports in one year 17,000,000 tons of earthy matter.* The river Witham, in Lincolnshire, before the recent improvements were carried out, passed through beds of shifting sands at its mouth. The tidal flow was stopped by a sluice across the river about eight miles above the mouth, and consequently the ebb was very sluggish when there were no land floods running down, the tidal water entering the confined portion of the river at the rate of from 3 to 4 miles an hour. During the dry summer of 1868, when there was no fresh water flow down the river, the amount of sand brought up and deposited along the bed of the channel was calculated to be one-and-a-half million tons.

* 'Irrigation in France,' Trans. Inst. C. E., vol. li.

the whole of which was removed and carried back to the outfall when the winter floods came.

Allowing $\frac{1}{700}$ as an amount that would be carried by a stream without overloading, this would be equal to about 0·09 lb. in every cubic foot of water.* Taking a main drain having 30 feet bottom, with slopes of 2 to 1, depth of water 8·0, velocity 1½ mile an hour, the quantity of earthy matter carried in suspension would be 117 tons an hour, as follows :—

The area is 368 feet, velocity 132 feet per minute;

$$\frac{368 \times 132 \times \cdot 09 \times 60}{2240} = 117 \cdot 1 \text{ tons an hour.}$$

Allowing ten hours for a working day, 1171 tons of earth, if loosened and broken up in the form of mud, would be carried away by the water.

MACHINES USED FOR DREDGING, SCOURING, ETC.—As a practical illustration of the working of this system the dredger employed by the Deeping Fen Trustees, hereafter described, was employed in cleaning out the Vernatts drain, which receives the water pumped from Deeping Fen, in Lincolnshire, containing 30,000 acres. The velocity of the stream, where the dredger was at work, varied according to the state of the tide in the river Welland, being very sluggish at high water, and increasing to about 1½ mile an hour at low water. The boat was employed on a section 170 chains in length, for eleven weeks, and during this time the whole of the weeds and mud accumulated, together with a portion of the bottom of the drain, consisting of clay, in places very hard, was broken up by the dredger, and transported by the water free and clear, not only of the drain itself, but also of the channel of the river, and deposited in the estuary ten

---

* Allowing 7000 grains in 1 lb. avoirdupois, and that a gallon of water weighs 10 lbs., this would give 70,000 grains in a gallon. Taking the proportion of $\frac{1}{700}$ would give 100 grains of earthy matter to a gallon of water. Or taking the cubic foot of water at 62·5 lbs., and the same proportion would give $\frac{62 \cdot 5}{700}$ = 0·08928 &c. lb. in a cubic foot of water.

miles distant. The total cost of working the boat, for labour, coal, oil, &c., was 65*l.*, equal to about 7*s.* 6*d.* per chain. It was estimated by Mr. Harrison, the surveyor of the district, that to have done this work by spade labour would have cost 200*l.*

In the river Welland, by the aid of this machine, a length of 24 chains was deepened 2 feet for a width of 17 feet in three days' working. In the river Glen the channel was deepened 3 feet 6 inches, with a very slow current running, the soil being stiff clay.

Several appliances for breaking up shoals and loosening the bed of streams have been brought out and used at different times, both in this country and abroad.

In the river Stour a boat was used, fitted with a kind of rake at the bow, adjustable to the depth required. At the side of the boat wings were fixed which could be extended so as to form a temporary dam when the rake was lowered. This dam caused the water to rise on the upper side. As soon as a head of from six inches to a foot had accumulated, the pressure forced the machine forward, dragging the rake along the bottom. The rate of progress was about three miles an hour. The machine is said to have been very effective in scouring the river.*

In removing the shoals in the river Maas a boat was used, fitted on each side with a screw propeller, 3 feet 6 inches in diameter, which could be raised or lowered by gearing on the deck. The screws were adjusted so as nearly to touch the shoals, and as the boat moved, these were caused to revolve at the rate of 150 revolutions per minute, by bevel gearing, driven from a cross-shaft by belting from the fly-wheel of a portable 12 horse-power engine. By the aid of this machine, shoals consisting of sand, clay, and peat, were easily and quickly moved, sand at the rate of 130 cubic yards an hour, clay and sand mixed, 116 yards. The boat was moved to

* Trans. Inst. C. E., vol. ii.

the shoal to be removed by the action of its own propeller, and was then anchored and warped along the shoal. This machine was afterwards converted into a suction dredger, having a two-bladed propeller or fan 4 feet 6 inches in diameter, which was brought halfway up the stern in a case, and the sand pumped up, discharged into the river on the ebb tide, and so carried away.

The navigation of the Danube in Hungary, being frequently impeded by shoals caused by the deposit driven out of the tributaries in times of flood, the Danube Steam Navigation Company employed a boat, fitted with a triangular rake 18 feet long, having thirty-four teeth 12 inches deep. This rake was hung over the bow of a steamer and dragged across the shallows, the steamer running astern, it being found that if the rake was hung over the stern it interfered too much with the steering. By the aid of this rake, the shoals, consisting principally of shingle, were removed, and the depth of water increased from 3 to 4 feet.*

For breaking up the sand on Pluckington Bank in the Mersey, to facilitate its removal by the scouring sluices, various forms of rakes and dredgers were used, a description and illustration of which will be found in the Trans. Inst. C. E., vol. xc.

A scouring dam, intended for the removal of shoals consisting of sand, was brought out by Mr. John Kingston, and was described and illustrated in 'Engineering' of August 4th, 1882. This machine, which has been successfully used for clearing the entrances to the tidal and coast canals at Balasore and other places in India, consists of a barge, over the stern of which is a framework, carrying a movable dam made of wood, with appliances for raising or lowering the same. As soon as this sheet dam is lowered into the water across the stream, and within a few inches of the bottom, the water is held up on the upper side of the dam and the current forced

* Trans. Inst. C. E., vol. lx. p. 387.

Drainage by Gravitation.

underneath, causing the sand or mud to be stirred up and mixed with the water. The force of the head of water is utilised to move the boat forward and drag along the bottom a "hedgehog," so as to loosen the hard crust. The machine is allowed to drift slowly down with the ebb current, and the material, thus broken up and scoured out, is carried away in suspension. Care has to be taken in using this machine in rivers passing through sand, that the water does not find its way round the sides and so scour holes in the banks.

The machine already referred to as being in use in the river Welland and the Deeping Fen drains, has been designed and brought into practical use by Mr. Alfred Harrison, superintendent of the Deeping Fen Drainage District. It consists of a barge, to which, from framework projections at each end, is suspended a "hedgehog" or revolving drum, on the periphery of which are spearheaded spades. The barge is moved along the channel by means of steel ropes, anchored in the bank at one end, and the other working round drums on the boats similar to those used for steam ploughing apparatus. The drums are made to revolve by gearing attached to a semi-portable engine in the boat, the one drum uncoiling and the other coiling up the rope. The drums are balanced by chains, passing over pulleys to counterbalance weights, so as to enable them to rise over any substance too hard for the spades to penetrate, undue strain on the ropes being thus prevented. The barge travels at the rate of about two miles an hour. The framework to which the hedgehogs are attached can move laterally by means of a handle, and thus acts as a steering apparatus, by means of which the boat travels round very sharp bends without difficulty. The spades are placed alternately, and only enter a short distance into the bed of the stream. By the constant travel of the two hedgehogs up and down the drain, the soil is broken up sufficiently small for the whole of it to float and be carried away by the water. A perpetual churning motion is carried on by the fore and aft rollers, and the water being breasted up by the fore-roller, causes a thorough mixture of the soil with the water, the earth being converted into sludge. Although soil and sand can easily be removed, a greater effect is obtained with a clay bed from the lighter specific gravity of this material. An illustration and description of this machine will be found in the 'Engineer' of October 28th, 1887.

For removing mud shoals which collected along the Mare Island Straits at San Francisco, revolving buckets were used, the machinery being actuated by the forces of the stream. A floating framework was fitted with an undershot wheel 20 feet in diameter. This wheel was driven by the current, and was geared into two drums, which carried double ropes fitted with small buckets. The ropes passed round loose pulleys, which dropped into the mud, carrying the buckets with them, and these on their return discharged their contents into the stream, by which it was carried away.

PROPORTION OF SIZE OF DRAINS TO LAND DRAINED.—No fixed rule can be laid down for the designing of a system of drainage for any particular district, either as to the number of the drains, their arrangement, width, or depth, as these must all be governed by the particular conditions of the land to be drained and the outfall.

A typical case may, however, be taken, having an area of flat land containing 20,000 acres, discharging into a tidal stream which only allows the sluice doors to be open fourteen hours out of the twenty-four. The rainfall is assumed as being that due to a quarter of an inch in twenty-four hours, the whole of which is to be discharged off the land. The quantity of water due to this rainfall would be equal to 210·11 cubic feet a second; the main drain, running only 7 hours each tide, would require a capacity equal to 359·498, say 360 cubic feet a second. Taking the district as $6\frac{1}{4}$ miles long by 5 miles wide, this would require one main drain

Drainage by Gravitation. 25

up the centre, with lateral drains on each side at intervals of every fifty chains; ditches branching out of these lateral drains would be required at intervals of every ten chains, leaving the areas of the fields drained by them twenty-five acres each. Taking the main drain as having 26 feet bottom at its lower end, with side slopes of 2 to 1, a mean depth in floods of 8 feet, and a rise in the bottom of 4 inches per mile, and 2 inches per mile on the surface, its discharging capacity would be as follows:—

	feet.
Bottom width	26
Width surface of water	58
Mean width	42
Mean depth	8

Area $42 \cdot 0 \times 8 \cdot 0 \quad = 336 \cdot 0$
Contour $17 \cdot 6 + 17 \cdot 6 + 26 \cdot 0 = 61$ $\quad \frac{336 \cdot 0}{61} = 5 \cdot 508$ H.M.D.

Fall per mile $0 \cdot 166 \times 2 = 0 \cdot 332$

$\sqrt{5 \cdot 508 \times 0 \cdot 332} = 1 \cdot 352$ vel. per second.

$1 \cdot 352 \times 0 \cdot 80 = 1 \cdot 0816$ „ „

Area $336 \cdot 0 \times 1 \cdot 0816 = 363 \cdot 42$ cubic feet second.

The dimensions in the drain at the middle would be 40 feet 6 inches top, 10 feet 6 inches bottom, 7 feet 6 inches depth, and discharging capacity 182 cubic feet a second. The quantity of drainage due to the rainfall on 10,000 acres, the quantity discharging at this part of the drain would be $180 \cdot 06$ cubic feet per second. At the top end the drain would diminish to $1 \cdot 0$ bottom, $29 \cdot 0$ top, and 7 feet depth. The twenty lateral drains would each take the drainage of 1000 acres. The mean dimensions of these drains in the middle of their length would be 2 feet 6 inches bottom, 6 feet 8 inches top, slopes $1\frac{1}{4}$ to 1, mean depth 1 foot 8 inches, fall in surface 6 inches in mile, and in bottom 1 foot in mile, discharging capacity $5 \cdot 19$ cubic feet per second, the constant for friction, &c., being taken at $0 \cdot 70$. The water due to the rainfall of

500 acres, the quantity coming in at the middle of the lateral drains 5.25 cubic feet per second. These drains are taken as running the whole 24 hours. The field ditches to have 1·0 bottom, slopes $\frac{3}{4}$ to 1, mean depth of running water 1 foot. Supposing that in floods, during the time the sluice doors were closed by the tide, the water rose 1 foot above the mean level, as before given, and fell at low water to 1 foot below this level, the drains would hold, between low and high water, a quantity equal to the rainfall of five hours. Supposing, in anticipation of a flood, and before the water had swollen in the outfall, the water in the main drain was run off within 3 feet 6 inches of the bottom, the drains would hold up to 3 feet below the surface of the land a quantity equal to about one day's rainfall. The area of land occupied by these drains would be 282 acres, equal to $\frac{1}{7\frac{1}{1}}$ of the whole area. If the water had to be pumped, and the pumps worked night and day, the area of the drains would be proportionately less, as would also be the case if the sluice was situated sufficiently above the outfall to allow of a longer discharge than 14 hours.

The sluice for this district would require to have three openings of 14 feet each. The water would approach with a velocity of 1·078 feet a second, and as the piers of this sluice would have pointed ends, it may be assumed that it would pass through at the same rate, the velocity of approach being sufficient to overcome the friction due to the sides and bottom of the openings. The depth of the water on the sill being 8 feet, the same as that in the main drain, the area of the waterway would be $3 \times 14\cdot0 \times 8\cdot0 = 336\cdot0 \times 1\cdot0816 = 363\cdot42$ cubic feet per second.

SLUICES.—Low fen and marsh lands being almost invariably below the level of the tides, their outfall drains must be protected at the point where they discharge into the sea or tidal stream by doors constructed to keep out the tide when it rises above the level of the water in the drain.

These doors are generally made self-acting, so that they automatically open on a falling tide, as soon as the water in the river is slightly below that in the drain, and close as soon as the tide rises above the level of the outflowing water. A very small head is sufficient to open and close these doors.

In large drains the doors are made in pairs, the size of the opening for the pair seldom exceeding 20 feet in width. The doors being self-acting, a greater width than this is unadvisable, the concussion of the doors as they come together on closing with the rapidly increasing head of the rising tide throwing a considerable strain both on the frame-work of the door and the masonry.

The doors shut against a solid wooden sill at the bottom and a pointed frame at the top, hooded over. In some sluices the doors are made of sufficient height to be above the rise of the highest tide. This plan adds considerably to their weight and cost, and from the great length renders them liable to strain without any corresponding advantage. The angle at which the doors are set is generally about 25 degrees; it should not be less than 20, or exceed 30 degrees. The heel post works in a masonry hollow quoin on a pivot at the bottom, and is held in place by an anchor-strap at the top. These doors do not open back flush with the wall, as in a lock, but stand off sufficiently to allow the rising tide to get behind and act on the back so as to close them. In order to give stability to the structure, the masonry opening is made of considerable width, and forms a bridge over the drain.

The length of the piers against which the doors open is about double the width of the opening, and they are invariably, as are also those on the inside, made with pointed or rounded ends. On the inside of the arch draw-doors are also placed, working in grooves, and made to lift up or down by means of a cast-iron toothed bar, actuated by pinions and gearing so adapted that one man has complete control over the door. In large doors counterbalance weights are suspended by chains over pulleys. These doors are kept open to their full extent in floods, and lowered when the flood is over, so as to regulate the water in the drain to a certain fixed height for the purposes of navigation, or for the supply of water to the ditches for fencing or other purposes. These drains being frequently used for barges, one of the openings is in this case provided with lock doors for the admission of craft when the doors are not open.

Smaller sluices, with openings not exceeding 4 to 5 feet, are made either with a single door, hung in the same manner as already described, or with "tankard lid" doors, the door being hung by a pair of double-acting hinges from the top, and frequently provided with a counterbalance weight attached to a lever fastened to the door.

An improved method of hanging flap doors has been designed by Mr. Stoney, by the use of which this description of sluice doors may be used of considerable size. Attached to the flap or door are a pair of lever bars, having counterbalance weights at the upper end. These bars are carried on the segment of a circle, the centre of which is situated in the centre of gravity of the moving mass. This segment rocks on a horizontal to the path situated above the water-level, making the friction less than when fixed pivots are used, and rendering the opening and closing very easy. Doors of this description were used for the sluices of the Ballyteigne and Kilmore Reclamation Works, a description and illustration being given in the 'Engineer' of April 29, 1887. These doors, 6 feet 3 inches wide by 4 feet 6 inches high, were found to open with a head of from $\frac{1}{4}$ to $\frac{3}{8}$ inch.

In addition to the sluice doors at the outlet, it is frequently desirable to place doors at the end of drains discharging into the main outfalls, to prevent backing up of the water in heavy floods and at tide time, when the water accumulates, especially where the main drain has to receive water from high land. By the use of these doors a large amount of embanking may be saved, and even where embankments already exist the erection of doors saves great pressure, and the risk of a breach, a contingency from which no banks are free. Some settlement, or weak place in construction, or burrow made by mole, rat, or rabbit, which may have been in existence for some time unknown, is finally discovered by a flood a few inches higher than usual, the water finds its way through, the bank bursts, and a whole level is inundated.

Great care is required in making the foundations of sluices to prevent the water from finding its way under the floor at times when, owing to the head on the outside during high tides, the power of the water to penetrate through the ground is very great. The lands enclosed generally consisting of deposits of alluvial matter, and the site of the sluice being on the seashore or on a river, the soil is frequently nothing but silt or sand, and frequently no material of a more tenacious character can be reached within a distance that would warrant the foundations being carried down to the solid stratum. When the soil consists entirely of silt, the danger to guard against is that of the water finding its way under the floor of the apron and invert of the culvert. Several cases have come under the author's experience where this has occurred, and the sluices, otherwise well built, have been left standing on the bearing piles with the material entirely washed away from under the concrete below the floor. The only reliable course to pursue in dealing with foundations in sand or silt is to build the sluice on bearing piles, and completely to box in the whole of the site covered by the foundation with sheet piling, driven eight or ten feet below the bed of the channel into which the sluice discharges. Within the box a solid bed of cement concrete to be placed, and on this the planking for the floor. Wings should be carried out with box piling for some distance each side, both at the inner and outer ends, to prevent the water making its way round outside. The brick-work of the piers and for the culvert will rest on the planking, bearing piles being driven to carry the walls; the whole upper part being surrounded with puddled clay. In fixing the depth to which the piles are driven, consideration must be given to the probability of the deepening of the out-fall, either by improvements or natural scour. After the sluice has been built cement grout should be forced under the floor, by pumping it in through holes bored in the planking, and afterwards plugged. By this means every cavity becomes filled up.

The illustrations given are examples of a small sluice with 8 feet opening built on a silt foundation, and of a larger sluice with three openings of 14 feet each, suitable for the draining of a district of about 20,000 acres.*

In situations where a solid foundation can be obtained the difficulties of construction are less. The following is the description of the foundations for the Ferraby sluice for the Ancholme drainage as given by Sir J. Rennie.† The sluice consists of three openings, each 18 feet wide, and one lock 20 feet wide. The sill is 2 feet 6 inches below L.-W. spring tides in the Humber. The soil is alluvial silt and clay. Beech or elm piles, 24 feet long and 12 inches diameter, were driven 3 feet apart, centre to centre, all over the site of the foundation. The earth was then excavated 2 feet below the pileheads, and blocks of chalk were well rammed in and grouted with lime and sand. Fir cap-sills and transoms, 12 inches by 12 inches, were fixed on the top of the piles, the space between these being filled with brick-work in Roman cement, and the whole covered with Baltic fir three inches thick, fastened down with 9-inch jagged spikes. The flooring was well bedded with lime, puzzolana, and sand. On this floor inverted arches of stone, 18 inches deep at the crown, were built, and the piers and sills erected. On the Humber side an apron was made with piles 16 feet long, the spaces between being filled in with blocks of chalk to a depth of 3 feet, and covered with 3-inch planking.

* Plate 1, Figs. 1, 2, 3, 4, for the small sluice, and Plate 2, Figs. 1, 2, 3, 4, for the large sluice.
† Trans. Inst. C. E., vol. iv.

Drainage by Gravitation.   31

In selecting the site for a sluice discharging on the sea coast, as sheltered a position as practicable should be chosen, where the sand is least likely to drift and fill up the outfall. Shifting sands are frequently a source of great trouble, and where sluices discharge on sandy fore-shores, it often becomes necessary to carry the outfall from the sluice for a considerable distance by means of a covered wooden tunnel.

When the coast is flat the water, after leaving the sluice, frequently has to travel through a long stretch of marsh and sandy foreshore. Where creeks already exist, it is generally necessary to deepen and straighten them. To prevent the shifting of the channel through the sand, and the sides from being washed down in storms, fascines made of thorns are the best materials that can be employed. These, when once properly laid in their places, and bedded with the sand, form inexpensive and efficient training walls, which with slight care will last for a great number of years. The same method has also been successfully applied to the training and straightening of rivers. A full account of the method of fascine training channels through sandy estuaries will be found in a paper by the author in the 'Transactions of the Institution of Civil Engineers.'[*]

Where the sluice discharges into a river, the site should be fixed at a concave bend, where the water is always deeper than in other parts, and the sill less likely to be blocked by deposit in dry weather. The position of the sluice should be such that the direction of the outflowing water should join the ebb current at as small an angle as practicable and so that its direction should coincide with that of the river, so as to cause as little disturbance or check as possible at the point where the two streams meet.

Where the district to be drained is at a low level, the sill of the sluice has generally to be placed at the lowest level at which the river is ever likely to be scoured out or deepened.

[*] 'Fascine Work at the Outfall of Tidal Rivers,' vol. xlvi.

It may even be of advantage, when every inch of fall is of consequence, to place it a certain depth below the bed of the river. Although the bed of the river at the time the sluice is built may be below the sill, the out-flowing water will always keep the doors clear, the water at the bottom of the drain rising up to the higher level in the river, as it does over a sunken weir. Any deposit that may accumulate against the doors in summer is soon washed away when the winter rains cause the water in the drain to rise. For the purpose of scouring this deposit away, small sluice doors are placed in the larger doors, near the bottom. These being drawn by means of rods worked by gearing, permit a current of water to pass out at low water, having sufficient head to wash away the deposit so as to allow the main doors to open.

The quantity of water that will pass through a sluice is found by multiplying the area of the water-way by the velocity, the dimensions being taken at the outer side of the sluice. The full velocity due to the head not being acquired until the outer side of the sluice is reached, if the depth be taken on the inside, it would give too great a result. The velocity is governed by the head, less the frictional resistance of the sides and the disturbance caused by the eddying of the water as it leaves the confined space between the walls for the open stream. If the piers have rounded and pointed ends, the frictional resistance will not amount to 4 per cent. With square piers and rough masonry, the loss due to the friction will amount to 15 or 20 per cent. In sluices with doors which do not open freely, such as in structures with hanging doors, or with lifting doors which are not raised freely out of the water, there is resistance on four sides, and the discharge in such cases may not amount to more than 60 per cent. of that due to the head. The amount to be allowed for frictional resistance must be a matter of judgment determined by the facility for the outflow of the water with the least possible friction or disturbance by projections causing

Drainage by Gravitation. 33

eddies. The constant to be used in the formula for calculating the discharge, varies from ·96 in the best form, and large sluices to ·60 in small structures with hanging doors. In calculating the velocity, allowance has to be made for the head due to the velocity with which the water approaches the sluice. If the sluice is of sufficient capacity, this will overcome the friction, and carry the water through without any heading up.

For example, take the case of a sluice with three openings of 16 feet each, and a depth of 9 feet, the drain running with a velocity of 1·50 feet per second, and discharging 1200 cubic feet per second. The quantity passing through each opening would be 400 feet, and the velocity required with the area of the sluice would be 2·778 feet per second. The head required for this velocity, allowing 4 per cent. for friction, that is multiplying by the factor ·96, would be ·1308 feet. Deducting the head due to the velocity of approach of 1·50 feet or ·0381, leaves the head required at the sluice ·0927 feet, or a little over one inch.

Area of waterway = 16·0 × 9·0 × 3 = 432·0 feet.

Total quantity coming down drain divided by area of sluice = $\frac{1200}{432}$ = 2·778 velocity required.

$$\text{Head required for this velocity} = \left(\frac{2 \cdot 778}{8 \times \cdot 96}\right)^2 = \cdot 1308 \text{ Feet.}$$

$$\text{Less head of approach} = \left(\frac{1 \cdot 50}{8 \times \cdot 96}\right) = \cdot 0381$$

$$\text{Head required at sluice} \quad \ldots \quad \ldots \quad \ldots \quad \cdot 0927$$

The proportion of depth to total width of the water passing through sluices, varies in practice from one-fourth in the smaller sluices to one-fifth and one-sixth in those of larger capacity.

Taking the average of four sluices draining pure fen land on the East Coast, with total width of openings varying from 33 to 45 feet, and a mean width of 37 feet, the number of acres to each foot of waterway is 1082. Allowing the depth of water to be one-fifth of the width or 7·40 feet, and that the water passes through with a velocity of 1·50 feet per second, this nearly allows for the discharge of rainfall due to quarter of an inch in 24 hours over the area drained. Of six large sluices draining rivers and mixed fen and high land, having openings varying from 74 feet and a mean of 55 feet, the number of acres to each foot of waterway is 2304. Allowing the depth to be one-fifth of the width or 11 feet, it would require a velocity of 2·19 feet per second for sufficient water to pass through to allow for a rainfall of a quarter of an inch in 24 hours. The discharge, however, being due partly to high land, the quantity of water passing through these sluices is less than this quantity.

SYPHONS.—Syphon pipes can be used for the discharge of water from enclosed lands into the outfall instead of sluices, but the cost of working them, and the loss of head required to carry the water through, places them at a disadvantage as compared with the ordinary sluices.

When an accident occurred to the Middle Level sluice on the River Ouse in 1862, it became necessary to place a solid dam of a very substantial character across the drain, and in order to afford means of discharging the water from the drain into the river, syphons were erected under the direction of Sir John Hawkshaw.

The syphons erected at the Middle Level were 16 in number, laid across the dam at an inclination of 2 to 1 on either side, each end being terminated by a horizontal length containing the upper and lower valves. The upper surface of the lower pipes was laid 1 foot 6 inches below low water of spring tides, and the top of the syphon was 20 feet above the same level. The syphons were of cast iron $1\frac{1}{8}$ inch thick, 150 feet in total

Drainage by Gravitation.

length, and 3 feet 6 inches in diameter. They were put into action by exhausting the air from the inside by an air-pump worked by a 10 horse-power steam engine. These syphons continued in use for 15 years. Owing to their capacity not being sufficient to cope with heavy floods and to discharge the water with sufficient rapidity, there was frequently a difference of more than 4 feet between the level of the water in the drain and that in the river, the average varying from 2 to 3 feet, a very serious loss in such a flat district. It being found that the cost of adding a sufficient number of syphons to drain the fens effectually would be greater than that of building a new sluice, Sir J. Hawkshaw reluctantly advised the latter course, although contending that the syphons were right in principle and practice.

CHAPTER III.

FIELD DRAINAGE.

THE object of field drainage is, by facilitating the discharge of the surplus water to such a depth from the surface, that while, on the one hand, it is removed so far from the roots of the plants as not injuriously to affect them, yet, on the other, it is not so deep as to retard, during the dry weather of the summer months, the supply of moisture which will arise from the substratum by the action of capillary attraction. The rain which falls in summer time is nearly all absorbed by the vegetation and the dry soil, or is evaporated, and little or none of it soaks through the ground to the ditches. In winter, however, about 60 per cent. of the rainfall soaks through the ground, and is carried away by the drains to the ditches and outfalls. Supposing that rain has been falling for some time, that the ground has become thoroughly moist, and that the pores of the earth are full of water, the rain then percolates through the interstitial spaces, and by the law of gravity proceeds downwards, until its progress is arrested by some impermeable stratum or soil already fully charged with water. It then accumulates, rising higher and higher, until it arrives at a line level with the water in the river or main drain of the district. This level is termed in the fens the "soc" or "soak."

In properly drained ground, while the rains of winter leave the surface soil in a healthy moist condition, that below the drains becomes completely saturated; and this supply of moisture is gradually drawn up, by capillary action, to supply the loss of moisture in the upper soil, which in dry weather is absorbed by the roots of the plants or evaporated by the summer suns. During the severe drought of 1887, Mr. T. F. Hunt, Assistant Professor of Agriculture at the University of Illinois, made some experiments in order to ascertain the comparative quantities of moisture in drained and undrained soils. The average of forty samples of earth, 2 feet in depth, taken from undrained fields was 13·2 per cent., while in those taken from drained land it was 14·1 per cent. Probably the draining had been done at a small depth, or the difference would have been greater. The small difference in favour of the drained land, however, may serve to dispose of the objection to draining, that it renders land less able to bear drought.

Thus it will be seen that, other considerations apart, pipe-drains should be laid sufficiently deep to remove the surplus water from the roots of the plants, yet not so deep as to retard the moisture from rising, when wanted, from the supply stored up in the stratum below the drains.

Drainage also acts mechanically on a tenacious soil, and assists in the discharge of the rainfall and the improvement of the texture of the ground by contracting it, and thus increasing the number and size of the larger pores, making more numerous crevices. That this is the case may easily be proved by taking a roll of wet clay, 1 foot in length, and drying it, when it will be found to shrink in length about half an inch, which, in a drain 100 feet long, would be equal to increased spaces which, if added together, would measure 4 feet 2 inches. The value of these crevices and contractions may be more fully realised by examining the appearance of two seeds of corn, the one of which has been sown in well-drained land and the other in a hard, cold soil. In the former case, the rootlets are able to travel in all directions in search of food, and the plant is strong and healthy; in the latter the delicate fibres of the roots are unable to force their way through the hard ground, and the plant, lacking nourishment, is stunted and unhealthy.

The admission of air to the soil not only improves its texture, but also raises the temperature, and supplies nourishment to the plants. There is no doubt that a well-drained and consequently a well-aërated soil, requires much less manure than one that is sodden with water. There are many mineral and organic substances in all soils which remain dormant and useless to vegetation until decomposed by the action of the atmosphere; there are also many salts which are unaffected by the water in the ground, but which, on exposure to the air, are immediately set free and dissolved, and carried to the roots of the plants. An excess of water will thus neutralise the chemical decomposition of the substances contained in the manure laid on the fields, and which largely supply food to vegetation. Drainage is as useful in promoting the circulation of atmospheric air as in removing the superabundance of moisture; for if the pores in the soil are emptied of water, it is evident that their place must be supplied with air; and as the effect of drainage is, by mechanically improving the texture of the soil, to increase the number of these crevices, so it also increases the circulation of the air, which passes through the soil to the drains, and along them to their outlets, thus keeping up a constant supply of fresh air.

TEMPERATURE OF DRAINED LAND.—The temperature of the atmosphere attains its maximum on an average of seasons about the middle of July—the cold period attaining its maximum about the middle of January. The heat that is given out in the summer is absorbed by the earth, and gradually finds its way downwards until it reaches a depth beyond which, speaking generally, the temperature of the soil is not affected by the heat of summer or the cold of winter. This depth is found to vary from 50 feet to 100 feet below the surface, the variation of temperature between winter and summer being only 3 degrees at 24 feet below the surface, the mean variation of the atmosphere being, on the surface, nearly

30 degrees. The heat travels through the soil at a rate proportionate to the depth, as will be seen from the following table:—

Situation of Thermometer.	Middle of Warm Period.	Middle of Cold Period.	Mean Range.
	Month	Month.	Degrees.
In the air	July 21	January 20 ..	29·8
Sunk 1 inch in ground	July 26	January 24 ..	25·4
,, 3 feet ,, ...	August 9	February 8 ..	21·7
,, 6 ,, ,, ...	August 25 ..	February 24 ..	15·4
,, 12 ,, ,, ...	September 25..	March 27 ..	9·5
,, 24 ,, ,, ...	November 30 ..	June 1	3·4

Thus it will be seen that it takes six months for the alternations of heat and cold to affect the soil at a depth of 24 feet; and when it is coldest above ground, the subsoil at this depth below the ground is the warmest, and the heat of the summer sun is gradually ascending through the soil during the winter and early spring months to assist the germination of the seeds sown, and to keep warm the roots of the plants during the snows and frosts of winter.*

The effect of judicious drainage is to increase the capacity of the soil for absorbing heat, and also to enable it to keep up the temperature of the soil during cold weather.

Water is a better conductor of heat than air, and thus in cold weather, and when the ground is covered with snow, undrained land, having the crevices or spaces between its particles filled with water instead of air, on the one hand parts with its supply of heat more rapidly than drained land; and, on the other hand, is less calculated to take in as large a supply in the warm period of the year.

To prove the effect of drainage in raising the temperature of the earth, a premium was offered by the Marquis of Tweeddale, some years ago, for observations and experiments to be made on soils of a similar character, growing the same crops, and situated in the same locality; the result of which

* Steinmetz, 'Sunshine and Showers.'

was a collection of carefully prepared and thoroughly reliable observations, from which the following results are culled. That during a long-continued frost the mean temperature of drained land at 30 in. below the surface was nearly 1½ degree warmer than the undrained. That showers of sleet and cold rains lowered the temperature of drained lands 2 degrees, and undrained land 4 degrees. That in every instance drainage gave a decided advantage in an increase of temperature, except only in summer, when a heavy fall of rain was found to lower the temperature of the drained land 1 degree more than the undrained—an evident advantage to a hot, parched soil.

Experiments also made by Dr. Madden led him to the conclusion that an excess of water in the soil reduced its temperature in summer 6½ degrees, which amount he considered equivalent to an elevation above the level of the sea of 1959 feet. So that, supposing two fields, lying side by side, the one drained, the other undrained, and supposing them both equally well cultivated, there would be nearly as much difference in the amount and value of their respective crops as if the drained one was situated at the level of the sea and the other on an elevation as high as the Pentland Hills.* Dr. Madden also, in order to dispel the idea where it existed, that the interstitial spaces being so minute that their contents could be of no consequence, quotes the fact that in moderately well-pulverised soil they amount to no less than one-fourth of the whole bulk of the soil itself; for example, 100 cubic inches of moist soil contain no less than 25 cubic inches of air. According to this calculation, in a field pulverised to the depth of 8 in., every acre will retain beneath its surface no less than 12,545,280 cubic inches ; and for every extra inch in depth the ground is cultivated, 235 tons of additional soil are called into activity, and rendered capable of retaining beneath its surface 1,568,160 additional cubic inches of air.

* Lecture on Agricultural Science by Dr. Madden.

Undrained ground is less calculated to take in a store of heat in summer than drained land, the summer sun being wasted in drying up, by evaporation, the winter rain from the soil, and in the process cooling down the land.

INCREASED VALUE OF LAND FROM DRAINAGE.—The actual increased return from drainage must vary a great deal, according to circumstances and the nature of the soil. Numerous cases were given in evidence before parliamentary committees of rents being raised 50 to 100 per cent. after drainage. The average of several different classes of soils showed a net return of 10 per cent. on the outlay. As a fair average it may be taken that on clay soils a wheat crop will yield one quarter to the acre more on drained than on undrained land, and this without any additional seed or labour. In wet cold seasons the increase will be much greater, and the drainage is often paid for by the extra produce of a single season.

After a series of years the subsoil of a thoroughly drained field changes into the nature of soil as far down as the level of the water in the drains, due to the ameliorating effects of air and water producing healthy decomposition of the organic and inorganic constituents.

When the working of the land and the treading of the horses is considered—a treading which in the case of a pair of horses leaves more than 200,000 footprints when cutting a 9-in. furrow over an acre of land—and the effect of this in puddling a wet clay soil and injuring its texture, the advantage of freeing such a soil from surplus water may more fully be estimated. Also by the penetration of roots and by their ultimate decay in the subsoil, and by the working of earthworms, the texture of the soil is improved. The drainer has not a better assistant than the worm. Worms work their way down through dry soil to great depths. The author has seen worm-holes at depths of 10 feet and 12 feet below the surface; and in a drained soil their burrows always extend as low as the pipes, the cavities made in their progress acting most effectually as feeders to the drains.

Thoroughly drained fields stand wet and *drought* better than undrained fields of the same sort of soil. From the principles already laid down, it is evident that this should be the case. It is well known by those who have paid attention to the matter, how during protracted droughts the thoroughly drained fields call attention to themselves by their superior verdure. By their improved texture they are not liable to become baked, and the free soil is in a condition to take in a supply of moisture from the dews of the summer night, which the hard dry skin of the undrained land is incapable of doing.

Thoroughly drained fields are also more easily tilled, and are in a fit state for the operation of tillage a much greater number of days in a year.

TIME TO DRAIN.—The time of the year chosen for putting in drain-pipes must be regulated by the cropping and other circumstances; but it may be stated that the drier the weather the better for the drainage. In clay soils, the drying action of the air and wind on the trenches allows the soil to contract and form the crevices necessary for the rains to escape to the drains. Experienced drainers recommend the month of February for the work, and that the pipes receiving a light covering of soil, should be left open through March, if it be drying weather, by which means the cracking of the soil is much accelerated, and the complete action of the drains advanced a full season.

In laying drains in a silty soil, the worst time to choose is when the ground is full of water; the feet of the men working in the grips cause the silt to purge, so that it is impossible to get a good and even bed to lay the pipes on; and even when laid they are extremely liable to choke, by the loose silt in the trenches being washed in by the water which pours out of the ground. If the pipes are laid when the silt is dry, or only slightly wet, the bottom of the trenches may then be taken out hard and firm ; and the ground, owing to the effect of the drainage, will never again be so charged with water as to make them liable to be stopped up ; and even should the ground, from any exceptional cause, be drowned, after the soil in the trenches had once become settled and consolidated, there would be no danger of the water washing it into the pipes, as it would find its way to them through the regular crevices or canals.

DEPTH.—With regard to the depth at which pipe-drains should be laid, this must depend to a great extent on the outfall. There may be special circumstances where drains may be advantageously laid at great depths, and by neglecting to descend a few inches in certain soils, many of the benefits of drainage may be lost. Again, where springs occur it is often necessary to lay the drains at considerable depths, and to resort to special means of getting rid of the water, as by boring down through an impervious soil to the porous stratum below. But these are cases which seldom occur in low land drainage. There are many cases, however, in fens and marsh land, where the state of the outfall ditches will not allow of a greater depth than 2 feet. The pipes *should never be laid so low* that their ends are buried in the water in the ditches into which they empty; such a practice is simply laying pipes for the purpose of soddening the land with water instead of draining it. It completely stops the whole circulation of air, and arrests all the benefits to be derived from a properly laid drain. A drain laid 2 feet deep, and free at the end, is far more effective than one laid 3 feet with the outfall constantly under water.

The outfall of the drains where they empty into the ditches should be constantly inspected to see that they are free and not stopped with weeds and earth. The ditches ought to be regularly scoured out and cleaned once at least every season. It is a good plan to lay the last tile of the main drain on a flat paving tile or brick, and to place a small iron grating before the mouth of it to prevent the vermin from getting up; and too much care cannot be bestowed in keeping these outfalls clean and free.

DIRECTION AND FALL.—Drains should always be laid to run with the fall of the land, and not across it. A different theory was held for a short time by some drainers, but practice has proved what theory would teach, that the drains should fall with the land, the only exception being in the case of springs. A consideration of the subject will show that the water has the least distance to travel to the drains when laid in this manner, and when there will get away most quickly. Supposing the strata to have the same inclination as the surface, and the drains to be laid 30 feet apart, the water will of necessity flow in the direction of the strata, and a part of it must therefore travel 30 feet if the drains be laid to run across the slope; but, on the other hand, if they be laid to run with the inclination, the water will flow from the centre space between the drains in both directions, and thus have only 15 feet to travel, or only half the distance.

Where a field is in one plane, and level throughout, it is better to lay the main across the centre of the field, letting the drains radiate from it at right angles towards the sides. The object to be kept in view is that the drains may be placed so deep, that while rapidly collecting and conveying away the surplus of the rainfall, they shall also be so situated as most effectively to promote a circulation of air, and allow the moisture to be drawn up from the subsoil below the drains to the roots of the plants in dry weather. Where no special circumstances arise to prevent it, this object seems to be most effectively attained where there is a covering of 3 feet on the top of the drain pipes; and this may be taken as a safe depth to lay drains, whether in tenacious clays or silts and more porous soils; and from 8 yards to 9 yards apart in the former class of soils, and from 10 yards to 12 yards in the latter, is sufficient distance for the drains to act effectively.

The object being to get rid of the water quickly, the less run it has through the small pipes the more rapid will be the discharge, the friction in the mains being much less than in the smaller drains. And so, in a very large field, it is never desirable to lay the smaller drains of a greater length than 200 yards. Some engineers allow 300 yards as a maximum length, and instances have come under the author's observation where 2-inch pipes laid in a clay soil in lengths of 20 chains have been in effective working order for the past ten years, and will possibly remain so as long as the pipes last; but under ordinary circumstances, 10 chains may be taken as the maximum safe working length.

In flat districts it is seldom that much fall can be given to the pipes, but a very slight inclination will cause the water to travel with sufficient rapidity to the outfall. It is better that the drains should be laid perfectly level than that a fall should be acquired by laying the pipes shallow at one end and deep at the other, the advantage gained by a fall thus acquired being neutralized by the varying effect the difference of depth must have on the uniform drying of the ground. The distance the drains are apart is determined with reference to the depth; therefore if the drain be laid shallow at the upper and deeper at the lower end, the distances must either be too great at one end or too little at the other.

Fall is not absolutely necessary to the safe working of drains. By the action of gravity, water is attracted towards the earth's centre, and travels towards that point until its progress is arrested by some impediment. Water varies from more solid substances in that all its particles are free to act, and they have so little cohesion, that every particle is free to obey the influence of gravity, and seeks the lowest place it can find; a natural fall is thus caused on the surface of the water in the drains. A fall in the drain itself only assists this action, because all falling bodies acquire a velocity in proportion to the height from which they fall; and so the greater the fall the greater the velocity, and the greater the velocity the greater the power to overcome obstacles, and the more certain and rapid the discharge. Thus, while fall in the bottom of the drain is a great advantage, and even a necessity, in drains which convey water having matters in suspension, as in town sewers, in enabling the water to keep the drain free from deposit, it is not absolutely necessary to the discharge of clear water, or for land drains laid at a proper depth ; and many miles of drains have been laid that are now doing their work well which have not an inch of fall.

Where pipes are laid level, or where only a very slight fall can be obtained, the main should, wherever possible, be laid lower than the small drains, and the end pipes should always tip, or be laid at a greater inclination than the others, in order to assist in drawing off the water.

The connections of the drains should never be made at a right angle, but the smaller pipes ought always to be made to enter the mains with a curve, or at a very obtuse angle. When one current of water impinges on another at a right angle, it causes a stoppage in both, and hinders the flow : an eddy is thus created, and any heavy matter held in suspension is precipitated, having a tendency to choke the pipes; whereas, if the smaller stream has the same direction given to it as the larger, by the pipes being made to join the others with a curve, the united currents flow on together without interruption. Experiments made in sewer work resulted in ascertaining the fact that, when equal quantities of water were running direct, at the rate of 90 seconds ; with a turn at right angles, the discharge was only effected in 140 seconds; whilst with a turn or junction, with a gentle curve, the discharge was effected in 100 seconds.

Irregularities in the cutting of the trenches and in the form of the pipes are far more injurious to efficient drainage than want of fall. It is of the first importance that there should be no hills and holes in the bottom of the trenches, but that a

Field Drainage. 47

regular inclination, when there is a fall, should be given throughout the whole length of the drain. In selecting pipes, those should be chosen which are evenly burnt, and which are not warped or twisted, and care be taken in laying them in the ground that the ends properly fit.

Men by constant practice acquire a wonderful skill in judging of the fall of the ground, and the regularity they give to their trenches ; and where the ground is wet and the water either runs away or follows them, they cannot get far wrong ; but in dry ground, and especially where it is uneven, the eye of the most practised drainer is apt to be deceived. Too much attention cannot be bestowed to this part of the work, and the pipes should never be laid in the trenches or covered up until the work has been inspected by a trustworthy foreman. To lay out a large system of drainage, the use of a spirit-level is absolutely necessary. For the drainage of single fields, the ordinary "boning rods" commonly used by workmen in setting out short sections of earthwork are sufficient. These rods are made in the shape of the letter T, about 3 feet 6 inches long, the cross being 14 inches, and the size of the wood 2½ inches by ½ inch. One of the rods has a leg with the feet and inches marked on it, which leg is made to slide up and down by means of two screws working in a slot, and can be fixed at whatever depth the drain is to be cut. The rods should be painted white, and to render them more visible the top of one should have a black line about half an inch deep on its upper edge.

The method of using is as follows :—A peg is driven in the ground at the upper end of the trench, and another at the lower end. The level of these pegs being set by the spirit-level, with reference to the outfall, the drainer causes one of the rods to be placed on the upper peg and the other on the lower, the rod, with the adjustable leg being set to the depth the trench is intended to be, is moved along it so as to keep the three in a true line. Any elevation or depression in the bottom of the trench is by this means at once detected. The use of these rods is acquired with very little experience, and levels can be ascertained with quite sufficient accuracy for all practical purposes.

If it is desired to find the inclination or fall of the ground, all that is necessary is to fix two pegs about 10 feet apart from each other, making them level with the aid of a straight-edge and spirit-level, or with a carpenter's level and plumb-bob; and then holding the two boning rods as before on these pegs, the third rod with the sliding leg is to be held at the lower end of the trench, or wherever else it is required to level to, and then sliding out the leg until the tops of the three rods are in a line; the distance the leg has to be drawn out gives the fall.

PIPES.—After trying various sizes and shapes for the pipes, opinion is now universally in favour of cylindrical tubes, 2 inches in diameter and 1 foot long, for the small or feed drains, and from 3 inches to 4 inches in diameter for the mains. Some tile-burners manufacture a circular pipe, having a flat bottom; if they could ensure that these would burn without the least twisting, there would perhaps then be a slight advantage in their having a better bearing on the bottom of the trench; but as this is never the case, the flat bottom is worse than useless, in rendering the pipes heavy and cumbersome. The author has repeatedly watched men laying these pipes, and half were not laid with the flat part downwards, the reason given being that the men could not make the ends fit when so laid. The circular pipes are less liable to warp and bend in the burning, having the same thickness of material on every side, and are therefore easier and better to lay. Collars are occasionally used, but are quite unnecessary, except in very rotten ground, when they are useful in assisting to keep the ends of the pipes from dropping away from one another. In such ground, in order to lay the drains effectually, the expedient should be resorted

Field Drainage.

to of putting sods at the bottom of the trench, and treading them well down, so as to give a firm bed for the pipes to lay on. This is often absolutely necessary, and the only way of putting pipe-drains in boggy soils.

As the expense of carting pipes from the maker's is a consideration in the cost, it may be mentioned that a one-horse cart will carry 800 2-inch or 500 3-inch pipes; and one horse will take this load easily on a good road, but it will require two horses to drag it over soft ground.

COST —The cost depends upon so many local circumstances—as the quality of the soil, the rate of wages, the depth at which the pipes are laid, and the distances apart—that it is impossible to give any fixed or definite sum. But it may be stated, as an average, that two men can dig out the trenches in a soft clay soil free from stones, lay the pipes, and fill in again at the rate of from four to five chains a day; and that an average price for pipes, at the maker's yard, is 21s. per 1000 for 2-inch pipes, and 42s. for 3-inch pipes. Having ascertained the cost of the pipes, and the rate of wages for the district, the cost per acre can be calculated from the following table :—

Distance apart.	No. of Pipes required for One Acre.	No. of Chains of Digging.	
Yards.		Chains.	Rods.
5	2,905	44	0
5¼	2,640	40	0
6	2,420	38	2¼
7	2,075	31	1
8	1,816	27	2
9	1,613	24	1¾
10	1,452	22	0
11	1,320	20	0
12	1,209	18	1⅓
13	1,117	17	0
14	1,037	15	3
15	974	15	0¼
16	907	13	3
16½	880	14	1⅓

NOTE.—The contents of this chapter are extracted from a pamphlet published by the author in 1868, entitled 'Practical Remarks on the Drainage of Land,' copies of which may be obtained from the publishers of this book.

CHAPTER IV.

DRAINAGE BY STEAM POWER.

MACHINES FOR RAISING WATER.—The machine for efficiently draining low lands is one that will readily adapt itself to the varying amount of work to be done, owing to increase or decrease of lift from the rise and fall of the tide, or of floods in the outfall into which it discharges, and from the lowering of the water in the feeding drain as pumping proceeds. The parts should be as simple as possible, and the machine should be so constructed as not to get out of order from lying by. Owing to the intermittent character of the work, most pumps are idle for the greater part of the year.

Setting aside special contrivances which have occasionally been used, but, owing to their unsuitability, the use of which has not been repeated, the machines used for raising water for the drainage of land are scoop wheels, screw pumps, bucket pumps, and centrifugal pumps. Of these the scoop wheel and Archimedean screw pump are the oldest types of machine. The former still does duty to a greater extent than any other method of raising water for drainage purposes, although it is gradually being superseded by the centrifugal pump. Archimedean screw pumps have not been adopted in this country, but in Holland they have been largely employed.

Bucket pumps have been used in some instances, both in Holland and England, for land drainage, notably for the drainage of Lake Haarlem, a full description of which will be found later on. Bucket pumps are still in use for the drainage of the Waldersea district on the Nene and of the Marton district on the Trent. The use of these pumps was probably advised by engineers whose experience was acquired in mining districts, where most excellent results were obtained from pumps of the bucket type. Trials of bucket pumps have given out as useful an effect as any others here mentioned; but, as the pumps were designed for working at much higher lifts than those required for land drainage purposes, these trials do not afford a guide, the proportion of efficiency more rapidly diminishing as the lift decreases than in centrifugals. From the construction of these pumps, they are not adapted for a varying lift; and in cases where they have been applied to fen drainage, the water has always to be lifted higher than it need be, that at Marton raising the water as much as 6 feet higher than necessary. The valves and working parts are also ill-adapted to cope with water charged with mud and grit, and the weeds and pieces of wood which frequently find their way to the inlet.

SCOOP WHEELS *versus* CENTRIFUGAL PUMPS.—The question as to whether the scoop wheel or centrifugal pump is the better machine for draining land has been much debated, and the matter is still a subject of controversy. The older class of fen enginemen and managers place implicit faith in the scoop wheel, and believe it to be superior to all other machines. When, however, wheels have been replaced by efficient pumps the result has been so satisfactory that the author has never met with an engineman who would wish to return to his scoop wheel. Such instances have occurred, and the pump been removed and replaced by a wheel, but only where the pumps were of the most inefficient character and improperly driven. The pump being a machine of superior character needs more intelligence on the part of the person in charge, and, as with all other machines, requires care and skill in the driving.

This question was some time ago referred by the Dutch Government to a Commission, with instructions to report as to the best machine for raising a given quantity of water—in this case 140 tons a minute — to a height varying between

11·3 feet and 12·3 feet, and also at a height varying from 4·9 feet to 13·1 feet. To the first question the Commissioners were not able to give an opinion as to whether one form of pump was superior to all others for a high but nearly constant lift. The answer to the second question was decisively in favour of centrifugal pumps, as they found that no other machine applied itself so well to differences of level in the external and internal water. No other machine permitted the application upon so large a scale of the whole disposable motive force to all lifts comprised within the limits stated; and thus while the machine adapted for a maximum lift will with lower lifts discharge larger volumes, the useful effect which is produced by the coal consumed does not vary to any great extent. They therefore recommended centrifugal pumps for both kinds of work.

Subsequently, in 1877, Signor Cuppari, an Italian engineer, spent a considerable time in Holland visiting the different pumping stations and investigating this subject. The conclusion he finally arrived at was that no general rule can be given as to the employment of one or other of the different machines, but that all the circumstances of each case must be considered before a decision is come to as to what machine to use.* That the general opinion of Dutch authorities was that in choosing a machine, consideration should be given to the following circumstances, and the machine chosen which met these requirements best: the turbidity of the water; the probability of the internal water level being permanently lowered; the nature of the foundations; the method of establishing communication between the inner and outer water level; the level at which the machine can be placed with reference to the water to be discharged; the cost of erecting and working. That the centrifugal had the advantage in all these cases, except the first, over all other machines. That scoop wheels

* Cuppari 'On Water Raising Machines.' Trans. Inst. C. E., vol. lxxv. 1883-84.

Drainage by Steam Power. 53

are efficient machines and the best where there is a large amount of *débris*, and that they have the further advantage that they can be easily repaired by ordinary workmen. The motors for moving them may be of common types, but cannot be used to the best advantage, owing to the difficulty of adapting them to the slow velocity required for the wheel. That they further labour under the disadvantage, as compared to centrifugals, of requiring stronger foundations; with a high lift the wheel must have a large diameter, the sill must have a low level, and this necessitates massive and deep masonry. That when there is a liability of a permanent lowering of the low-water level, wheels would require costly alteration, whereas with centrifugals additional lengths can always be added to the piping, and the only difference is that the consumption of steam will be greater. That in regard to the separation between internal and external water, the easiest and safest arrangement is that of pumps which discharge the water through pipes carried over the banks or inserted in masonry walls of sufficient thickness, thus avoiding the sluices which are required for wheels or screws. That the system of direct action between engine and pump is one that is most economical in fuel; and that the centrifugal pump lends itself most readily for action with this kind of motor.

As regards expense, Signor Cuppari gives a table showing the cost of the pumping stations in Holland during the previous ten years, from which it appears that the average cost per horse-power of water lifted is as follows:—

	Building.	Machinery.	Total.
	£	£	£
Scoop wheels	46·14	46·28	92·42
Screw pumps	..	..	94
Centrifugal pumps	34·20	36·80	71
Piston pumps	..	..	72

Statistics given of the drainage stations erected in the seven years 1875–81 show that centrifugal pumps are steadily making their way in Holland, as out of 139 machines put up, 50 were centrifugal machines, 38 were scoop wheels, 30 screw pumps, 4 piston pumps, and the others of various types. Out of the 57 machines erected in recent years by Messrs. De Witt, engineers, of Amsterdam, 31 were centrifugal pumps, 23 were Archimedean screws, and 3 were scoop wheels. The centrifugal pumps generally used in Holland are direct acting, having horizontal spindles, the discs placed above the water level. The turbine form has been tried, but the results were not favourable.

With regard to the relative merits of scoop wheels and centrifugal pumps in the quantity of coal consumed, the general weight of opinion amongst engineers in this country who have had an opportunity of comparing the relative merits of the two machines is decidedly in favour of the centrifugal pump. This question was thoroughly investigated about ten years ago by Mr. J. M. Heathcote, of Conington Castle, a gentleman who was not only the owner of land drained by steam power, but was greatly interested in fen drainage. As the result of his investigations, Mr. Heathcote came to the conclusion that the pump was decidedly the more economical machine, and in this he was supported by facts and figures from other sources furnished by Messrs. Easton and Anderson. These, however, while useful so far as they went, were not drawn from actual trials of the two machines working under precisely similar circumstances. The nearest approach to this is the running of the two sets of machines over a series of years for the drainage of the Wexford Harbour reclamation, particulars of which will be given hereafter. For the three years 1881, 1882, 1883, the consumption of coal at these two pumping stations was about one-third in favour of the centrifugal pumps, or at the rate of 18·65 pence per acre for the land drained by a pump, and 26·30 pence for that drained by a scoop wheel. The latter was of modern construction, and the lift in each case the same.

Drainage by Steam Power. 55

To show that considerable difference of opinion still exists as to the comparative merits of scoop wheels and centrifugal pumps, attention may be drawn to the fact that recently in the same number of the 'Engineer,' an account was given in one part of the paper of the removal of centrifugal pumps which had been erected in Egypt by an English firm for lifting the water for irrigation purposes, and the substitution under the direction of a French engineer of scoop wheels; and in another part, of the intended removal of scoop wheels erected by Dutch makers for the drainage of land in the north of Italy, and a substitution of centrifugal pumps by English makers, because similar pumps working in the same neighbourhood had satisfied the authorities that they were the more efficient machines.

Having paid considerable attention to this subject, and had frequent opportunities of becoming acquainted with the working of both machines, the conclusion arrived at by the author is that, with regard to existing wheels, where a scoop wheel can be made efficient at a reasonable outlay, it would be more economical to adapt it than to replace it by a centrifugal pump. If the wheel requires replacing, or great expense has to be incurred in altering the masonry and foundations and lowering the wheel, it will be found more economical to replace it with a centrifugal pump. In all new drainage districts in this country, there can be no doubt that the direct acting centrifugal pump is the most efficient and economical machine to fix.

ENGINES USED FOR DRIVING WHEELS AND PUMPS.—In the early attempts to drain land by mechanical means wind power was entirely resorted to. In Holland, where this source of power is largely applied for driving machinery, the numerous windmills and pumps all over the country show that steam has not yet succeeded in displacing the more economical, if less efficient, source of power. Many of the old windmills driving scoop wheels still remain in use in the

Fenland. On the old one-inch Ordnance map there may be counted five hundred windmills in the Fenland, which are in use now, or were at one time engaged in draining the land. In the Littleport and Downham district, in Cambridgeshire, containing 28,000 acres, no less than seventy-five windmills were engaged in lifting the water off the land—work which is now much more efficiently done by two steam engines. In 1729 Captain Perry, an engineer who is known from his attempts to stop the breach in the banks of the Thames, erected a number of windmills for working scoop wheels for lifting the water out of Deeping Fen, in Lincolnshire. In 1824 the forty-four windmills which from time to time had been erected were replaced by one main pumping station, having two large scoop wheels driven by steam-power, and the Fen, which previously had been only in a half-cultivated condition, became completely reclaimed.*

Mr. Arthur Young, in his survey of Lincolnshire, made in 1799, gives the following description of a drainage windmill in use on the estate of Mr. Chaplin, in Blankney Fen :—" In that long reach of Fen which extends from Tatershal to Lincoln, a vast improvement by embanking and draining has been ten years effecting. . . . This is a vast work, which in the whole has drained, inclosed, and built and cultivated, between 20 and 30 square miles of country. Its produce before was little, letting for not more than 1s. 6d. an acre, now from 11s. to 17s. an acre. . . . This vast work is effected by a moderate embankment, and the erection of windmills for throwing out the superfluous water. The best of these was erected by Mr. Chaplin, of Blankney. The sails go 70 rounds, and it raises 60 tons of water every minute when in full work. The bucket wheels in the mills of Cambridgeshire are perpendicular without the mill; this, which is called *dritch*, has in it a sloping direction in an angle of about 40°, and within the mill. It raises water 4 feet. Two men are necessary in winter,

* 'The Fens of South Lincolnshire.'

Drainage by Steam Power. 57

working night and day at 10*s*. 6*d*. each a week with coals for a fire ; add the expense of repairs, grease, and all together will amount to 2*l*. per cent., with 1000*l*. first cost. Mr. Eckard, of Chelsea and Dover Street, was the engineer. It drains 1900 acres. Two years ago the floods overtopped the banks and it cleared the water out so quickly that not a single year was lost."

Previous to the complete drainage of Lake Haarlem, the rainfall from 75,357 acres of Polder Land was lifted into the low part, or Boezem, by two hundred and sixty-one large windmills, of an aggregate force of 1500-horse power. The complete drainage of this lake was subsequently effected by steam power, a full account of which will be given further on. In 1776 the first attempt was made in Holland at using steam power in place of windmills for drainage purposes. A pumping station and steam engine was erected at Arkelschendam, near Rotterdam, and another at Mijdrecht in 1792. These engines worked bucket pumps actuated by rods attached to chains which rode on an arc at the end of a large rocking beam, the piston-rod being similarly attached at the other end of the beam. As these engines consumed 31 lb. of coal per horse-power per hour, the use of steam did not extend much, as it was thought that steam could not be used economically. In 1825 steam power was used at Zuidplas to drive two Archimedean screws for lifting water a height of about 22 feet. These engines consumed 22 lb. of coal per actual horse-power per hour. This lake was emptied by the united action of thirty windmills working scoop wheels and the two steam engines in 1840, and subsequently kept dry. Each steam engine raised the water from 2352 acres to a height of 3·28 feet Each windmill with its scoop wheel raised the water from 1898 acres for the first half of the upper lift to a similar height, and from 2656 acres for the second half. The annual cost of maintaining these thirty mills amounted to 60*l*. per mill. These mills were superseded by steam in 1871.

58   *The Drainage of Fens and Low Lands.*

Steam power as applied to Fen drainage in this country came into use about sixty years ago. In 1820 Rennie applied one of Watt's engines to the working of a scoop wheel for draining Bottisham Fen, near Ely. The success of the steam engine in draining water from the Cornish mines naturally led to the adoption for land drainage of the same type of machines as used for that purpose. The engines were massive and substantial condensing beam engines, working at a steam pressure of from 3 lb. to 5 lb. The structure consisted of heavy cast-iron beams, working on girders resting on the walls, and supported by ornamental cast-iron columns. The connecting-rods were attached to crank shafts, on which were fixed pinions working into spur wheels, or into a toothed wheel running round the inside of the scoop wheel. The foundations for carrying these engines and for the bearings of the wheels, which often weighed as much as 50 tons, were necessarily of a heavy and expensive character, piling and planking having been almost universally resorted to for the foundations. These massive engines seemed in character with the ponderous scoop wheels which were then universally used for lifting the water. Most of those in the Fenland bear on their framing the name of the Butterley Iron Company as makers, and the excellency of the workmanship is shown by the fact that after running for more than half a century, the greater part are still worked, some almost without alteration, but others with only slight changes to adapt them to a more economical use of steam. The engines more recently erected have been of various types, the descriptions of some of the best of which will be found in the account of the drainage stations to be hereafter given.

Engines used for draining land should be of as simple a character as possible, and free from all complicated parts. The more nearly they approach the type of engines used for agricultural purposes, the less difficulty will be experienced in obtaining experienced enginemen. The saving of coal by the use of condensing engines, or of others in which complicated appliances are adopted for the same purpose, is not, except in the case of the largest stations, of so much importance in an engine used for the drainage of land as a saving in first cost of machinery and foundations, and for subsequent repairs.

The circumstances attending engines used in pumping mines or for the water supply of a town or similar purposes, and those used for the drainage of land, are so different that it is an utter mistake to take such engines as types. In the one case the engine is continuously at work with skilled mechanics at hand to carry out repairs and rapidly remedy defects. The saving in the cost of coal in such cases is far greater than the interest on the extra outlay for machinery and foundations, and forms the principal subject of consideration. Experienced men, equal to meet all ordinary contingencies, can be employed, whereas, in the event of a mishap with a drainage engine, a messenger has to be sent from some out-of-the-way place to a distant town to obtain the services of an engine-fitter. Most land drainage engines run for only a short period in the year—in dry seasons, perhaps, for only two or three weeks in a twelvemonth—the saving of coals as between a complex and a more simple machine does not therefore compare so favourably with the annual payment for interest on the extra cost of the more expensive machine. The fewer parts of a simple machine also reduce the risk of breakdowns, and drivers of the agricultural class have generally sufficient intelligence to deal, at least for a time, with such accidents as may happen. Drainage engines should therefore always be of the simplest type, but of the best workmanship and ample strength. The latter quality is one that should always be insisted on. The extra cost of the metal required in making a strong and substantial machine as compared to one beautifully finished but so lightly constructed that it is always shaking itself to pieces, forms so small a portion of the whole cost that it ought never to influence a maker.

In Holland the type of engine most generally in use is of the horizontal condensing single cylinder. In the older stations in the Fens beam engines have generally been adopted. In stations of more recent origin no one particular class of engine prevails. In Holland, France, and Italy, the direct acting horizontal compound engines have most generally been applied for driving centrifugal pumps. Vertical engines require less foundations and are less subject to strain if settlement of the foundations takes place, a common occurrence in spite of all precautions in fenland, than those of the horizontal type, yet although adopted in some few cases, do not appear to have received the attention they would seem to merit.

Where the area of land is small, say not exceeding 2000 acres, and therefore not sufficient to warrant the cost of brick buildings, the most economical arrangement is to use a semi-portable engine driving a centrifugal pump, the whole enclosed in a galvanised iron shed. The cost of foundations and erection of a chimney is thus avoided. The pump can be driven by belt or by direct gearing from the crank shaft.

THE MANAGEMENT OF DRAINAGE ENGINES.—Although the saving of coal as between one type of engine and another may not be of such consequence as in engines used for commercial purposes, yet the total consumption is a matter which ought to engage the most serious attention on the part of the managers, as on this principally will depend the annual cost of the pumping station and the amount of taxes required to meet the expenses. The fuel should bear a direct proportion to the amount of water lifted. If more than is necessary is used it is due to the fault of either the engineman, the engine, or the pump. The excess has to be paid for. As regards the first, too great caution cannot be exercised in selecting a steady, careful, and economical man. The best men can only be secured by paying good and sufficient wages. A good engineman may save his wages many times over by careful stoking : an incompetent man may not only run up the coal bill, but do irreparable damage to the machinery by ignorant management. The difference of the consumption of coal due to good and bad stoking is strikingly shown by the trials of enginemen at the agricultural shows. It may be assumed that the men who enter for these competitions consider themselves as superior to the ordinary drivers, or they would not enter for the competition. Selecting two of these competitions as samples, with an interval of ten years between, it will be seen that there was a marked improvement on the part of the men in the work done. Some portion of this may be due to the difference in the engines, but it would not amount to much; and it is fair to presume that the managers would take care that the engine provided for the trials should be a competent machine.

At the trials at the Lincolnshire Agricultural Show at Spalding in 1872, with an 8-horse-power portable engine, fifteen competitors entered the list. The best used coal at the rate of 7·86 lb. per horse-power per hour, the worst 20·2 lb., the average of the whole being 11½ lb., a difference of 61 per cent. between the best and the worst.

At Gainsborough in 1883 there were nineteen competitors. The best man ran the engine with a consumption of coal at the rate of 6·77 lb. per horse-power per hour. The worst used 8·95 lb. The average of the whole was 7·69. There was thus a difference of 2·18 lb. of coal per hour in the driving of this engine by picked men. Taking the ordinary type of drivers of agricultural engines, it may safely be taken that there would be a difference of at least 10 lb. of coal per horse-power per hour. With an engine running at 10-horse-power, this would amount to over a ton in 24 hours. Beyond this would be further waste in oil and damage to machinery by want of skill or carelessness.

With regard to the quantity of coal consumed, the Dutch engineers in their contracts generally stipulate that this shall not exceed 6·60 lb. of coal per horse-power per hour of water actually raised. Allowing an efficiency of ·55 for the machinery, this is equal to 3·63 lb. per I.H.P. Some of the best pumping engines for land drainage purposes in this country consume from 4 lb. to 4½ lb. of coal per indicated horse-power per hour, which is above the Dutch standard, being at the rate of 7¼ to 8¼ lb. per W.H.P. At the trials of the engines and pumps put up at Fos, Bouches de Rhone, the consumption of coal was at the rate of 4·45 lb. per W.H.P. or 2·47 lb. per I.H.P ; and in those erected for the drainage of the Ferrara marshes 4 lb. per W.H.P., and 2½ lb. per I.H.P.

Neither with pumps nor scoop wheels is the consumption of coal proportional to the lift, the relative quantity increasing as the lift decreases. This is what would naturally be expected, as the dead weight of the machinery bears a larger proportion to the total quantity of work the smaller the lift. At trials at the Halfweg station in Holland the consumption of coal varied from 14·20 lb. per actual horse-power per hour when the lift was under 1 foot to 5·5 lb. when it was doubled. With a lift of only 6 inches the consumption of coal was at the rate of 50 lb. per horse-power per hour, the difference being accounted for by the large amount of power required simply to drive the wheels. Mr. Barker, one of the commissioners, gives the consumption of coal for engines working centrifugal pumps, as varying with the lift, as follows—not counting coal for getting up steam :—

Feet of lift	1	2·40	3·30 to 5·30	7·20
Lbs. of coal per horse-power of water lifted	9·37	8·22	7·71	6·60

(Trans. Inst. C. E., vol. lxxv. p. 274.)

In recent contracts made for centrifugal pumps in Holland, this matter has been taken into consideration, and three conditions of lift are specified.

The conditions in some of the competitions for pumping machinery in Holland, are that on the completion of the contract for the erection of the machinery there should be two trials, one carried out by the enginemen of the contractors, and the other by those of the purchasers. Both trials to last over a considerable period. In the event of the consumption of coal exceeding that guaranteed by the contract, the contractor to pay three-fourths of the capital sum which would have to be paid by the purchasers to provide the additional coal.

INSURANCE.—A very considerable saving in the management of drainage engines might be effected if the Commissioners in charge of them were to avail themselves of the advantages to be derived from the regular inspection which is undertaken by the boiler insurance companies. The primary object of these companies is to insure the owners of boilers against the damage arising from explosion, but the greatest advantage derived is from the periodical inspection made by their agents, who furnish reports after each visit as to defects, whether arising from bad setting or from wear or corrosion. By the timely detection of defects, many explosions are averted, and the reports of the inspectors as to the general condition and management of the engines and boilers should prove a useful check on the manager, and will frequently be found to result in a considerable saving in the amount of coal consumed.

ENGINE DRAINS.—In designing pumping machinery for draining land, care must be taken that the power supplied is adequate to the work to be done. This will depend on the amount of rainfall in the particular district, and the proportion of it that has to be lifted in wet seasons. As every ton of water lifted represents the money value of the coal consumed in effecting this, it is obviously desirable that all high land water that can drain off by gravitation should be excluded from the drainage district by catchwater drains and banks. The pumping machinery should be adequate to the maximum rainfall of wet years, as it is at such times, when the outfall stream is full, that the benefit will be most felt.

The quantity of water due to rainfall has already been dealt with in the chapter on 'Drainage by Gravitation.' Assuming that the quantity to be raised to be that due to a quarter of an inch in 24 hours, this is equal to about 25¼ tons per acre, which, multiplied by the number of acres and the height to be lifted, gives the work to be done from which the actual horse-power of the engine required can be calculated. Thus, for example, taking a district of 1000 acres, with a lift of 5 feet, and daily rainfall of ¼ inch, this is equal to 39,388 lb. lifted 5 feet every minute, or 196,940 foot-pounds, which, divided by 33,000, the unit of 1-horse power, is equal to 6-horse power. Allowing that 50 per cent. is absorbed in overcoming the friction of the machinery and leakage of the pumps, 12 I.H.P. would be required. This is on the supposition that the engine during extreme floods is running night and day, which in cases of emergency is generally done. If the work is required to be done in less time, the power required will be proportionately larger. While it is desirable to provide adequate power, any unnecessary expenditure should be avoided as adding to the dead weight of capital on which interest must be paid. On the other hand, it is never desirable to put too much strain on an engine; a machine that is well master of the work will run more economically than one that is much pressed.

The best advantage is obtained when the machinery runs continuously night and day. If the drains are capable of delivering a full supply, a pump works at its best, and the scoops of a wheel being fully charged deliver their full quantity with less head than if the water-level is lowered and gathers again during the night. The coals used in getting up steam and restarting are saved, and usually a better result is obtained. If the machinery does not run at night, it will have to be so much larger as to render it capable of discharging the rainfall of 24 hours in 14 or 16 as the case may be, and the drains must be proportionately large to supply the increased capacity of the pump and wheel. The wages of the night men for the short time they are required will not be found to amount to as much in a large pumping station as the annual payment of interest and repayment of capital on the extra cost of the engines and drains required to run for only a limited time.

In calculating the work to be done, the height which the water is lifted is taken as the vertical distance between the surface of the water in the drain bringing the water to the pump, and that in the main drain or river into which it is discharged. If the water is discharged through horizontal pipes—as in one form of the centrifugal pump—an allowance has to be made for the friction. Consideration must be given to the fact that this height may vary considerably in the course of the day, owing to the outfall being a tidal stream, or from the water having lowered in the feeding drain during the time pumping is in operation.

If the pump used is of the turbine form, its position in the pump well should be sufficiently below the lowest surface to which the water is to be pumped to prevent its drawing air, as within reasonable limits the depth of the pump below the surface does not add appreciably to the work to be done. The covering over the pump should not be less than 2 feet. If the main interior drains are of sufficient capacity the inclination in the surface of the water should not exceed 3 inches per mile, and even 2 inches are sufficient to bring the water to the pump. The surface of the water in the drains for effectual drainage should be at a sufficient distance below the land to allow the drain pipes to discharge freely from the lands situated at the greatest distance from the pumps. In alluvial soils this will be from 3 feet to 3 feet 6 inches, and this, plus the amount to be allowed for the surface inclination from the furthest point, will regulate the level at which the water should be kept in the main drain.

The main drains leading to the engine, and especially the

main engine drain, act not only as conveyers of water to the pump, but also as reservoirs to collect the water when the engine is not running, and should therefore be larger than the calculation would warrant if merely founded on their discharging capacity. When steam is once up it is bad management to allow the engines to stand still for the water to gather, because the drains are not of sufficient capacity to keep them supplied. The pumping station should be placed as near the centre of the district as practicable. The main engine drain will then be of shorter length, and the minor drains arranged in a more effectual way as feeders than if the pumping station be fixed at one end of the district; the drains falling into it from both directions will also require less fall than if they had traversed the whole length. It is unnecessary to refer further to this matter, as the whole subject has been fully dealt with in the chapter on drainage.

COST OF PUMPING STATIONS.—The cost of erecting a pumping station depends upon so many circumstances peculiar to the locality that no definite figures can be given. Generally the cost may be taken for buildings and machinery at from 70*l.* to 80*l.* per horse-power of water lifted. Allowing for an annual average rainfall of from 25 inches to 30 inches, 1 horse-power would drain about 150 acres with a lift of 5 feet, making the cost per acre about 10*s.* The main elements to be taken into consideration are the quantity of water to be lifted—depending principally on the rainfall and the area drained, the height the water has to be lifted, and the nature of the foundations required. Frequently a much greater quantity of water has to be lifted than that due to rainfall from the soakage from other districts through badly constructed banks. If the district is small, the proportionate cost would be greater, but an increase in lift would be less in proportion, as adding little to the cost of buildings and general arrangements. The cost of erecting the Lade Bank engines in Lincolnshire, in 1868, was 17,000*l.*; the area of land drained, 35,000 acres. The pumps were calculated to raise 700 tons of water 4 feet 6 inches high per minute, equal to 80·37*l*. per horse-power of water lifted. The cost of the two stations at the Wexford Harbour Reclamation Works was 91·10*l*. for buildings and machinery at the north station, where the pump was erected, being 37*l*. for machinery and 54*l*. 10*s*. for buildings ; and 40*l*. per horse-power for the scoop wheel and engine—the cost of the buildings not being given in this station. The cost of an iron scoop wheel with curved blades and horizontal engine, erected by Messrs. Appleby, for the drainage of the Upwell district, in Norfolk, was 2680*l*.—equal to 74·68*l*. per horse-power of water lifted for the machinery, and 26·40*l*. for the buildings—together, 101·08*l*. Centrifugal pumps, driven by semi-portable engines, housed in wooden buildings, have been erected in the Fen districts at from 70*l*. to 80*l*. per horse-power of water lifted. In Holland the cost of erecting pumping machinery for horse-powers varying from 14 up to 500 during recent years has averaged 92*l*. per effective horse-power * for scoop wheels, the amounts varying from 58*l*. for the largest machines to 106*l*. for the smallest. The buildings have cost 46·1*l*. per horse-power, and the machines 46·3*l*. For centrifugal pumps the cost has been 36·8*l*. for machinery, and 34·2*l*. for buildings—together, 71*l*. The variation in cost for pumps has not been so great between the larger and smaller machines as that for wheels. The cost of erecting screw pumps is given at from 76*l*. to 100*l*. for buildings and machinery—the average being about 94*l*. per effective horse-power, the power varying from 12 to 130 horse-power.

In California, in the valley on the western side of the Sierra Nevada, where centrifugal pumps are used for raising the water for irrigation, and where a number of small separate pumping stations widely distributed are required instead of one large establishment, owing to the distance of transport,

* Cuppari, 'On Pumping Engines in Holland.' Trans. Inst. C.E., vol lxxv.

the machinery is designed of light sections and scant proportions. The cost of machinery for raising water to heights varying from 6 to 11 feet is given at 28*l*. 10*s*. per horse-power for machines capable of lifting about 30 tons per minute, and for larger engines up to 70 tons a minute 14*l*.*

COST OF MAINTENANCE.—The annual cost of maintaining a pumping station varies with the accessibility of the locality as affecting the price of coals, the efficiency of the machinery, and the skill and care of the engineman. From statistics prepared by the author as to the expense of maintaining pumping stations in a large district in the Bedford Level, the average cost for the three years 1881–2–3, which were very wet, and during which several floods occurred, was 16·25*d*. per acre throughout the level, or 1·86*d*. per acre per foot of lift, of which 1·47*d*. was for coals only. Taking the larger districts, in which, from the engines being of a better character, the proportionate consumption was less, the cost for coals only was found to be about 1*d*. per acre per foot of lift. These amounts were obtained as the result of the figures given by eleven different stations, draining about 120,000 acres of land, with lifts varying from 6 feet to 14 feet, the cost of coals being about 16*s*. per ton delivered.† During the same period, the cost of working the large engines and scoop wheels at Podehole for draining Deeping Fen in Lincolnshire, was 10·58*d*. per acre, of which 7·56*d*. was for coals. Taking the average lift of the water at 5 feet, this gives 1·51*d*. per acre per foot of lift for coals. The average working charges of the engines and centrifugal pumps at Lade Bank for draining the East Fen in the same county for the same period—1881–83—was 1089*l*., equal to 7·46*d*. per acre. The average rainfall was 30·27 inches a year. Taking the average lift at 4 feet, this is equal to 1·86*d*. per acre per foot of lift. The cost of coals at this station would probably be about 14*s*. per ton. For the

---

* 'Irrigation Machinery on the Pacific Coast,' by J. Richards.
† 'Report on the River Ouse.'

Drainage by Steam Power.

years 1871–72, in which the average rainfall was 28·25 inches, but was not so continuous nor the floods so high, the cost was only 532*l.*, equal to 3·63*d.* per acre ; the average lift being taken for that season at 3 feet 9 inches; this gives rather under 1*d.* per acre per foot of lift. At the Wexford Harbour Reclamation pumping station for the years 1881–83 the average rainfall was 40·34 inches, the average lift 5 feet 6 inches, coals 18*s.* per ton, the cost was 26·3*d.* per acre for coals for the scoop wheel, and 18·65*d.* for the centrifugal pump, respectively 4·76*d.* and 3·38*d.* per acre per foot of lift.

CHAPTER V.

THE SCOOP WHEEL.

THE "scoop" or "float" wheel has been in use as a machine for lifting water from very ancient times—it is mentioned by Vitruvius; and that the Romans used it for this purpose is proved by the discovery at the Tharsis Mines, in the south of Spain, a few years ago, of a scoop wheel which had been exhumed in the excavations then being carried on. This wheel was made of oak, of light scantling, put together with oak pins, no nails being used in the construction; and although it must have been underground for nearly two thousand years, the wood was in a good state of preservation.* The wheel was introduced into Holland by W. Wheler in a new form in 1649.

Mechanical power for drainage purposes in the Fen country came first into use about 200 years ago, when scoop wheels worked by horses were used by the Corporation of the Bedford Level. Horse-power was superseded by wind. In 1726 an Act was obtained for the drainage of Haddenham Fen by the use of windmills working scoop wheels, after which time their use became general throughout the Fen land, also for the drainage of the low land along the Trent and in other parts of the country. In Holland scoop wheels still largely exceed all other kinds of machines for lifting water from the low lands, and in Italy they are considered by some of the principal engineers as more effective for this special purpose than any machine yet invented.

These wheels have done exceedingly good service in the

* J. Lee Thomas, in Trans. Inst. C.E., vol. xxxii.

drainage of the land. When well constructed, and for situations where the height to which the water has to be raised is not great, and where there is not much variation in the lift, they are effective and useful machines, especially in the Colonies and remote places, where wood is more plentiful and available than a trained mechanic's services. The slow speed at which they travel fits them for being driven by windmills or the slow-speed beam engines by which these were succeeded. They are simple in construction, and easily repaired by the aid of such mechanical skill as is readily obtainable in country districts. They are not liable to get out of order when laid by, or easily damaged by floating substances brought to them in the water.

To the mind of those living by the side of the rivers and drains of low flat countries, and accustomed to the slow practices of an agricultural life, there is a sense of power and solidity about a massive beam engine with its slowly revolving fly-wheel and heavy beam rising and falling, driving a ponderous water wheel lifting a large mass of water, for which the small parts of a centrifugal pump and its rapid movements seem but a poor substitute. They are, however, exceedingly cumbrous, the wheel weighing as much or more than the total body of water lifted at each revolution. The larger wheels, of say 30 feet in diameter, weigh from 30 to 40 tons, and therefore require very heavy foundations and expensive masonry work for the wheel race. The slow-speed engines used for driving wheels are themselves as ponderous as the wheels, and also require heavy foundations and a large area of buildings. If engines of quick speed are used the loss of efficiency due to the gearing necessary to reduce the velocity of the engine to that of the wheel absorbs considerable power.

As generally constructed scoop wheels are very wasteful of power, and badly adapted to meet the alterations in the level of the water due to the falling of the level on the inside, as the water is pumped out of the drains, or on the outside due to the rise and fall of the tide, or of flood waters in non-tidal streams. As the angle at which the scoops enter and leave the water seriously affects the working of the wheel, even in the best constructed machines, there must always be a loss due to variation in the level. The unnecessary height to which some part of the water must be lifted also throws undue work on the engines, although it is not so regarded by all enginemen. Sir G. B. Airy mentions a case of a wheel which he visited, the attendant on which took considerable pride in his wheel because it lifted the water well up into the air, regardless of the fact that the steam power by which all this water was tossed in the air had to be paid for, and was so much waste of power.* There is also loss from leakage of the water between the wheel and the sides and bottom of the masonry trough in which the wheel revolves. In the event of the surface of the land drained lowering—a frequent occurrence—it becomes necessary to deepen the drains and lift the water from a lower level, involving the lowering of the wheel, or the lengthening the scoops, and reconstructing the masonry breast of the course, an expensive and difficult work, owing to the foundations being generally built upon piles.

Most of the old scoop wheels employed in the drainage of land are unnecessarily wasteful of power. Sir G. B. Airy, the late Astronomer-Royal, who examined one of these machines at work, for the drainage of a large district in the Fens, was of opinion that four-tenths of the power applied was wasted, a great part of which might have been saved by a proper arrangement of the parts, in which opinion he was confirmed by Sir W. Cubitt. Many of the wheels do not give off a useful effect of more than 30 per cent. of the power applied.

They are, however, in some case capable of improvement. By alterations recently effected in the engines and wheels used for the drainage of Deeping Fen, more than double the

* Trans. Inst. C.E., vol. xxxii.

The Scoop Wheel.

quantity of water was raised, the consumption of coals being at the same time reduced 42 per cent.

The scoop wheel resembles a breast-water wheel with reverse action. In its simplest form—Plate 3—it consists of an axle, upon which are fastened discs, to which are attached radial arms A, terminating at the other end in the rim B, upon which are fastened arms with boards called "scoops," C. The wheel revolves in a trough, connected with the drain on the one side and the river or place of discharge on the other. The scoops beat or lift the water from the lower to the upper side, the waterway on the river or outlet side being provided with a self-acting door E, which closes when the wheel stops. On the inside of the rim are cast-iron cogs fastened on in segments, and geared into these is a pinion keyed on to the shaft of the fly-wheel of the engine. In some wheels a spur wheel F is fixed on the shaft of the wheel, in place of the toothed segments. The former plan occupies less space, but the wheel is more difficult to repair in case of damage to the teeth. In some wheels the framework, including the rim, is made in four castings, two forming one side of the wheel, and are bolted together and keyed to the axle.

A large number of the scoop wheels in the Fenland were designed by Mr. J. Glynn, and made by the Butterley Company. These have a cellular casting or disc keyed on to the axle. The spokes are each in one casting through the width of the wheel, bolted through the disc and transversely to the rim. The wheels of more recent construction have had each spoke made in a separate casting bolted transversely to the disc and to lugs cast on the rim, the spokes being connected by struts and bolts. The rim is cast with sockets, in which are fixed with pins, oak arms, or "start posts." To the start posts are bolted boards, from 1 inch to $1\frac{1}{2}$ inch thick, varying at the circumference from 1 foot to 3 feet apart. Wheels of small width have only one start post, and the boards are placed lengthways parallel with the post. In wider wheels there are two or more start posts, the boards being placed in the opposite direction. These boards are called indifferently "scoops," "ladles," "floats," "paddles," the former term being adopted for use in this book, as that most generally in use.

The axle in the old wheels is carried on bearings resting on the masonry of the trough, the gudgeons having no means of adjustment, and are generally unnecessarily large, increasing the friction. In more modern wheels the gudgeons are kept in their place either by a shoulder running close against the plummer blocks or by screws bearing against the ends. The trough in which the wheel revolves is made of masonry, carried up as high as the centre of the wheel. The invert is made to the same radius as the wheel. The clearance, or space between the wheel and the trough at the sides and bottom, varies from $\frac{4}{10}$ inch in the best machines to $\frac{3}{4}$ inch, and even more, in the older ones. Owing to the large diameter of the wheels and their great weight, it is necessary that the foundation should be rigid and the wheel very nicely adjusted, as the slightest settlement causes the wheel to grind against the sides. Even with thorough adjustment there is always loss of water, owing to the space which must be left between the wheel and the masonry. In order to prevent this loss, some wheels have rims or shroudings on the sides, which partially or wholly enclose the water. The latter are termed wheel pumps. If only partially shrouded, only a portion of the leakage is stopped, and, if wholly, a difficulty arises in providing for the escape of the air contained in the space between the two scoops, which consequently do not become fully charged.

The scoops are generally flat boards bolted to the start posts and dripping from the radial line at an angle varying from 25 to 55, but generally about 40 degrees, the scoops being set out in tangents to a circle concentric with that of the wheel. The angle at which the scoop enters the water is termed the angle of ingress, being that which the face of the scoop makes with the surface of the water when first it comes in contact with it, the outer end of the scoop being at the apex of the angle. The angle of egress is that which the back of the scoop makes with the surface of the water when it leaves it, the end of the scoop as before being at the apex. In the diagram Plate 2, fig. 5, A B C is the angle of ingress, and D E F that of egress.

In determining the angle of the scoop the mean level of the water both on the interior and exterior side of the wheel must first be settled. The centre of the wheel is then fixed so as to divide the head and dip in equal proportions, except where the lift is great, when the dip is seldom made to exceed 6 feet.

Owing to the variation in both the external and internal water-levels, it is impossible to fix the scoops at such an angle as that they shall always enter and leave the water in the way most favourable to the discharge. The larger the angle which the scoop forms with the surface of the water the better the scoop enters, and the better hold it gets of the water; and the larger the angle of egress the better the water leaves the scoops; consequently the more the angle of ingress is improved the worse becomes the delivery. If the angle of ingress be too small, too much of the scoop comes in contact with the water on first entering, and instead of drawing it gently forward, beats it back, causing a disturbing element detrimental to the discharge. If the angle of egress be too small, the water does not drip off the scoop readily, but a portion is carried up with it above the level of the surface of the outfall, the height to which the part of the extra water is lifted in some wheels due to this cause being as much as 6 feet to 8 feet, while the remainder is lifted from 2 feet to 3 feet. The undue work thrown on the engine from this cause may be realised when it is considered that the lift of these wheels seldom exceeds 10 feet, and is frequently only half this.

Mr. Beijerinck, the engineer of several reclamations in

Holland, advises that the angles of ingress and egress of the scoops should be about equal with the mean internal and external water levels. If there is any variation, it is certainly better to increase the angle of egress, as this prevents useless lifting of the water at the point of discharge, and more is gained on the egress than is lost on the ingress side. Some of the Dutch wheels, however, are constructed so that the angle of egress is double that of the angle of ingress. The wheels at Katwijk, which are of recent construction, are made tangential to a circle concentric with the wheel, and having a radius of 5·25 feet, the diameter of the wheel being 29 feet 6 inches. This gives the angle of ingress with the water at mean level 20 degrees 30 minutes, and the angle of egress 42 degrees. The wheels at Podehole for the drainage of Deeping Fen, the larger of which is 30 feet in diameter, have scoops which incline from the radial line 25 degrees, being tangential to a circle having a radius of 3·75 feet. If the scoops were straight they would enter the water with 5 feet dip at an angle of 29 degrees, and leave it with an angle of 36 degrees. The scoops are 6 feet 6 inches long; the ends for a length of 18 inches being bent back from the straight line about 6 inches.

The best average results are probably obtained from wheels having an angle of ingress of 30 and of egress of 45, when the water is at its mean level.

In order to avoid the undue lifting of the water, and also to facilitate the entry, wheels, both in this country and in Holland, have been fitted with curved scoops made of sheet iron, the rest of the wheel also being generally of iron. At normal levels these wheels work well, but when the internal water level alters, the convex part of the scoop strikes the water, and any advantage otherwise gained is lost. Wheels with curved scoops should always be provided with a shuttle to adjust the flow of water. The variation in the level may, to a certain extent, be thus provided against. Curved wheels

The Scoop Wheel. 77

provided with a proper adjusting shuttle may be run with a greater speed than with flat scoops, as the water escapes off the scoops more readily, and will work effectively if run at a speed at the periphery of 9 feet per second. Curved wheels may also be made of less diameter than those having straight scoops. Wheels with curved scoops are in use at Zuidplas, Waterland, and other stations in Holland, for the Upwell and Outwell district in the river Ouse, and at Ravensfleet and Sturton on the Trent.

The older wheels made in Italy had the scoops straight and radial. These wheels were of low efficiency, dashing the water about as each scoop entered and left the water. More recently, the practice of some of the Italian engineers has been to make the wheels of iron, the scoops being inclined at about 60 degrees to the radius, and formed with a double curvature, a sliding iron shuttle being provided so as to admit the water to the lower part of the wheel only.* At Gouda, some of the scoop wheels erected in 1857 were subsequently changed to wheels having curves with the concavity towards the outer water, one having the concavity towards the inner water, and two with the scoops nearly flat. The difference in the delivery of these wheels was slight, the first-named giving, on the whole, the best result. These have since all been changed to flat scoops. As a matter of experience, wheels with flat scoops give the best results when all circumstances are taken into consideration.

The inlet and outlet courses for the water are constructed of masonry, the wall of the engine-house generally being placed on the inner side and the shafting from the engine passing through an opening. The practice in Holland is generally to decrease the width of the raceway as it approaches the centre of the wheel, the sides converging at an angle of 17 degrees from the straight line, the raceway at its widest part being about one-third wider than the

* Trans. Inst. C.E., vol. lxxvi., p. 402.

wheel. In some wheels, both in Holland and England, the divergence starts almost directly from the centre of the wheel, while in most cases the divergence begins at the point where the scoops are attached to the wheel. The object of this widening is to allow the water to get freely to the scoops both at the end and sides. The outlet channel is provided with either single or double self-acting doors, according to the size of the channel. These close automatically when the wheel stops. In some cases these doors are placed close to the wheel in a set-back in the masonry. A disturbance is thus caused in the flow of the water by the alteration in the size of the channel and by the water striking against the framing of the doors and the angles of the masonry. To ensure an even flow and prevent eddies, the outlet channel should have the same width as the wheel, and gradually widen outwards to the outfall drain; the sides should be smooth, and free from any projection or recesses; the door removed to some distance from the wheel; and the breast of the raised outlet sill rounded off.

Where the wheel has to contend with a great depth of water on the outside, the scoops as they come round, churn up a large quantity without discharging it. The wheel thus becomes choked, and does not part with its load as efficiently as it otherwise would. This churning up of the water causes also a strong back undercurrent, which seriously impedes the free discharge from the wheel. To show the effect of this, an instance may be quoted of a case where, although at the time a large body of water was being discharged by a wheel into the outfall channel, the undercurrent near the raceway was so strong that a small boat which had got loose near the wheel was sucked down and got jammed in the scoops. The attention of the superintendent being called to the circumstance, he caused the self-acting door at the end of the raceway to be cut in half horizontally, with the result that when the water in the outfall was high owing to floods, the lower half of

The Scoop Wheel. 79

the door always remained closed. To obviate this movable breasts have recently been fitted to some of the old wheels, which can be raised or lowered to suit the level of the water. The wheel at Podehole has thus been altered (see Plate 5). On the breast of the outlet sill an iron plate has been fitted in a recess cut in the masonry. This plate is hinged to another plate which lies on the floor of the outlet channel. By means of a segmental-toothed rack geared into a pinion on the windlass this movable breast can easily be raised or lowered, and the sill of the discharging channel adjusted to the height of the water. A similar arrangement has been carried out in the large wheel on the Hundred Foot River, and the discharge in both cases has been very greatly improved. Experiments made by the superintendent of the latter wheel showed that with the same pressure of steam in the boiler, and other circumstances being the same, the number of revolutions of the wheel—50 feet in diameter—increased about one-third, or from 31 in three minutes, with the movable breast lowered, to 41 in the same time when it was raised its full height of 4 feet, making a total rise of 8 feet.

In 1868 Mr. Samuel Naylor, the superintendent of the Morton Car drainage district on the Trent, took out a patent for a somewhat similar arrangement, which is thus described in the patent specification: "At the end of the channel or chase of the wheel is formed a face, and at the bottom thereof, in a horizontal axis, is arranged a flap or valve constructed so that it will float, and jointed at intervals. This flap floats in the water, allowing the water in the wheel to pass it in an upward direction. When the level of the water falls by the stopping of the wheel, and the flow tends towards the wheel, the flap floats up to its face and makes a tight joint, and so prevents the return of the water " —Patent specification, 11th March, 1868, No. 839. The difference between Mr. Naylor's plan and that already described is that the former is self-adjusting, and also supplies the place of the self-acting doors placed across the outlet. The wheel for draining the Morton Car district in the Trent has been fitted with this arrangement. Mr. Naylor also included in his patent a curved guide for causing the water to enter at the under side of the wheel, so that the scoops should not "come in contact with the water until they have ceased to descend, and are travelling horizontally, or nearly so, on the under side of the axis." The blades or arms in the wheel are curved backwards, or in the opposite direction to that in which the wheel revolves, so that the water may leave them advantageously on the rising side of the wheel.

The wheels at Katwijk, in Holland, have been fitted with a somewhat similar arrangement to that here described, the floor in front of the wheels being movable, and hung on hinges, so that it can rise up automatically, and form an adjustable breast, suitable to the varying level of the water in the outlet.

Very few of the old wheels have been fitted with shuttles to regulate the supply of water to the scoops. With a varying height of water it is not possible, without this arrangement, to work the wheel to the best advantage. By means of the shuttle the depth of immersion of the scoops can be controlled, and the wheel prevented from being overloaded. The shuttle is also useful in adjusting the load either at first starting, before the wheel has got into full swing, or, in the case of a tidal outfall, in diminishing the supply to the wheel as the height of lift increases with the rise of the tide. The shuttle consists of a wooden framework, with a sliding door fitted in the raceway close up to the wheel, sloping at such an angle as to be tangential to the circumference of the wheel. The door is provided with friction rollers, which work on a frame in the side next the wheel, and is provided with a balance weight and rack and pinion for moving and adjusting the shuttle. A shaft can be carried into the

The Scoop Wheel.

engine-house and geared to the pinion, so that the shuttle may be adjusted in the engine-room, a flange with float being also placed there to show the height of the water. See illustration of the Podehole wheel, Plate 5.

The diameter of the wheel is regulated by the "head and dip," that is, the distance from the lowest point of the ladles to the maximum height to which the water has to be lifted. Sig. Cuppari quotes a formula of Mr. Forster's for calculating the diameter $D = 9\cdot 82 \sqrt{i + p}$, in which i is the immersion of the scoops, or the dip, and p the lift $= 9\cdot 82 \sqrt{H}$, in which $H =$ the height from the lowest point of the wheel to the highest external level to which the water has to be raised, the measurements being in feet. The constant in the formula gives a larger diameter than is generally to be found in either the Dutch wheels or those in use in the Fens. A constant of $8\cdot 75$ gives a result more nearly approaching the Dutch and English practice. The wheels at Zuidplas, which have curved scoops, have a dip of $3\cdot 28$ feet, and a head of $11\cdot 8 = H$ $15\cdot 08$, with a diameter of $32\cdot 80$ feet, equal to about $8\cdot 5$ times the square root of H. The wheels at Katwijk, erected in 1880, have a head and dip of 11 feet, and a diameter of $29\cdot 50$ feet, equal to $8\cdot 92$. Taking twenty-five of the principal wheels in England, having an average extreme head and dip of $15\cdot 0$ feet, the mean diameter is $34\cdot 0$ feet, equal $8\cdot 77$ the square root of H. The new wheel at Wexford Harbour has a diameter of 40 feet for $14\cdot 50$ feet head and dip, equal to a constant of 10. Wheels with curved scoops require a less diameter than those with flat. The Italian wheels generally have a larger diameter in proportion to the lift than the Dutch or English. The largest wheels in Holland do not exceed 33 feet in diameter. In England the largest diameter is 50 feet, and several examples of wheels 35 feet and 40 feet exist. The limit for efficient working may be taken at 36 feet. It is not advantageous to use scoop wheels for lifts above 12 feet.

When the height is greater than this, centrifugal pumps are more effective and economical in working. Before the introduction of pumps, where the lift was great, it was usual to divide it into two, and to employ a double set of wheels, one working at some distance behind the other.

In 1872 a patent was taken out by Mr. G. Hamit, the superintendent of the Haddenham Drainage District, to meet the case of alterations required in existing wheels owing to the subsidence of the soil. By his proposal he claimed to save the expense of increasing the diameter of an existing wheel with the consequent lowering to the masonry of the trough. His invention is described as consisting of "the application of an auxiliary wheel at the entrance to the wheel race to feed the water to the scoop wheel, the efficiency of which is moreover increased. This auxiliary wheel is provided with curved blades and is rapidly rotated, the water being admitted to it through a sluice protected by a grating"—Patent specification, 3764, 1872. So far as the principle is concerned, there is nothing new in this, as examples of double lifts already existed in the Fens. The object sought could be obtained more efficiently by lengthening the scoops, as has been done in Deeping Fen and other places. The existing machinery of the old wheels being well and strongly constructed, is easily adapted to the increased work by simple alterations, and by using steam at a higher pressure. By judicious alterations to the engine and wheel a large saving of coals may be effected, and the engine rendered capable of dealing with the increased work.

The greatest quantity of water raised by scoop wheels, so far as the author's knowledge goes, is at Atfeh. At Katwijk, the six wheels raise each over 333 tons, or a total of 2000 tons a minute 4 feet high, or 1200 tons 7 feet high. The greatest quantity of water lifted at one station in England is at Deeping Fen, the larger wheel raising 300 tons a minute to a height of 5 feet, and the two 560 tons a minute. The wheel at the Hundred Foot River in Cambridgeshire discharges 120 tons a minute at a height of 17 feet, and 200 tons at the ordinary lift of 13 feet. The two Italian wheels at Adria discharge 300 tons a minute each, the maximum lift being 10 feet. The Egyptian wheels recently erected at Atfeh, on the Nile, can each discharge 254 tons a minute, or a total for the eight wheels of 2030 tons.

The speed at which wheels with flat scoops run should not exceed 8 feet per second at the periphery. When this is exceeded, too great an impetus is given to the water, and part of it is lifted higher than necessary. The best results are obtained with a slower rate of speed than this. The slower the speed the less the water is dashed about. It is, however, contended by some managers of wheels that a slow speed involves additional friction from the gearing required to reduce the speed of the engine to that of the wheel. In the old engines the number of revolutions of the engine to those of the wheel was 3 or 4 to 1. This has been increased to 6 or 7 to 1. A high speed, it is also contended, has the advantage of the head due to the velocity with which the water leaves the wheel, provided the outlet channel is of suitable form. Thus, with a speed of 8 feet per second, the water will pass through a regular and smooth waterway into an outfall, the surface of which is nearly 1 foot higher than the water at the wheels. Wheels in England having a diameter of 30 feet generally make 4 to $4\frac{1}{2}$ revolutions a minute, equal to a speed of 6·27 feet to 7 feet per second. Some of the Dutch wheels run at as low a speed as 3·46 feet per second. Four of the new wheels at Atfeh make 2·29 revolutions a minute, and the other four 1·91 revolutions, equal to a speed of 3·93 feet, and 2·95 feet respectively. Wheels having curved scoops can be run at a higher velocity than those with flat blades. Where the discharge is into a tidal stream, wheels are regulated to run at varying speeds according to the height of the water in the outfall, so as to adapt the power of the engine to the varying lift.

The work done by a scoop wheel is measured by the quantity of water lifted in a given time. To ascertain this it is necessary to know the head and dip of the scoops. The "dip" is the depth the scoops are immersed in the water when vertical in the trough. The "head" is the difference in level of the surface of the water on the inner and outer side of the wheel. The cubical quantity of water lifted each revolution is ascertained by multiplying the mean circumference of that portion of the wheel which is immersed by the width of the scoops and by the length of the immersed part. From this must be deducted the space occupied by the start posts and scoops. Mr. Wilfrid Airy, who paid considerable attention to the performances of scoop wheels, and published a pamphlet,* giving the result of his investigations into their working some years ago, considered that a deduction should also be made for leakage of the water which falls from the wheel in the "clearance," or space left between its sides and bottom and the masonry. He estimated this as equal to an amount due to the area of such clearance multiplied by the velocity due to the head or height the water was lifted. Thus, taking a wheel 30 feet diameter, with scoops 4 feet wide, and having a clearance of ½ inch, the lift being 5 feet and the dip 5 feet, the calculation for loss from this source would be as follows:—The area between the scoops and the masonry would be $5 \cdot 0 + 5 \cdot 0 + 4 \cdot 0 \times \frac{1}{2}$ inch $= \cdot 583$ feet. The velocity due to the head $8\sqrt{5} = 17 \cdot 89$ feet per second. The area multiplied by this velocity gives $625 \cdot 7$ cubic feet a minute, or nearly 11 per cent. of the whole quantity lifted. This is greater than is found to be the case in practice. If the theoretical velocity as found above be reduced 30 per cent., to allow for the friction of the water against the sides of the

---

* 'Remarks on the Construction of the Course, and Design for a New and Improved Scoop Wheel,' by W. Airy, 1870.

The Scoop Wheel.

masonry, it would bring the result more in accordance with the results which have been obtained from the best constructed wheels. Signor Cuppari states that the actual discharge of the wheels at Zuidplas, which have curved scoops, is 92 per cent. of the theoretical. The author has seen these wheels at work and does not consider their work as calculated to give out such a good result. Taking the dimensions as given above, the gross discharge of the wheel would be 5764 cubic feet per minute, found thus :—Diameter of the wheel 30 feet. Deduct half the dip of the scoops off each of the radii of the wheel, leaves the mean diameter of the immersed part of the wheel = 30 − 5 feet dip = 25 feet, of which the circumference is 78·54 feet. Area of the scoops 5·0 × 4·0 = 20 feet; 78·54 × 20·0 = 1570·80. Deducting the area occupied by the scoops and start posts, 129·80 feet, gives 1441 cubic feet for each revolution of the wheel. This multiplied by four, the number of revolutions, gives the discharge as 5764 cubic feet per minute. If the leakage be taken as that due to 70 per cent. of the theoretical quantity found above, say 438 feet, the net discharge would be 5326 cubic feet per minute. The lift being 5 feet, the horse-power in water lifted—W.H.P.—would be $\frac{5326 \times 5 \times 62·5}{33,000} = 50·43$.

The power required, in addition to that for lifting the water, in overcoming the frictional resistance of the machinery of the engine and wheel, and also for useless work in lifting water too high, varies considerably. Mr. Airy estimated the percentages of loss due to unnecessary lifting of the water, friction, and other causes as follows :—From leakage, 8 ; unnecessary lifting of the water, 19 ; friction on the gudgeons, 5 ; resistance from the shape of the course, 8 ; a total of 40 per cent. To this must be added 10 to 15 per cent. for the friction of the gearing, making a total loss for the wheel alone of 55 per cent. There is no doubt that in many of the old unimproved wheels the loss from the working of the engine and wheel amounts to as much as 70 per cent. of the power applied; on the other hand, in the best wheels, this has been reduced to 30 per cent. Messrs. Watt and Co., of the Soho Engineering Works, Birmingham, who have had considerable practical experience in the working of scoop wheels, having altered several of the older wheels in the Fenland, and thereby added largely to their efficiency, are of opinion that the merits of the scoop wheels are not sufficiently valued. Having carefully measured the quantity of water flowing down the feeding drains, and taken the power of the engine as indicated at the cylinder, they have found an efficiency of 75 to 80 per cent. in wheels with flat scoops, worked by beam engines using steam at a pressure of from 20 pounds to 25 pounds on the square inch. The details of one of these trials will be given in the description of the Hundred Foot wheel. Messrs. Watt consider that in a scoop wheel the flow being continuous from the feeding drain to the actual delivery, as good an effect should be obtained in a well-constructed wheel as with a reciprocating pump, where the motion of the water is not only changed in direction, but where frequently masses of water as between the pumps and air vessels or other pipes have to be put in motion. Their experience gained from fitting up a large number of waterworks, leads them to rely on obtaining with such pumps 80 per cent. of effective work. With regard to the loss by leakage, Messrs. Watt entirely disagree with Mr. Airy. They consider that in a well-made wheel the loss from this cause is practically nothing. The head that causes the leakage they consider is only that between the level of the water in one space between the scoops and the level in the next succeeding space, and that the general motion of the wheel and the upward current of the whole mass of water practically overcome the leakage. Further, they consider that in calculating the discharge of a wheel a greater quantity must be allowed for than the exact dip of the scoop, as owing to the velocity imparted to the water by the outer diameter of the scoops, combined with the gradual contraction of the feeding drain increasing the velocity of the water, the cavities between the scoops are filled from a fourth to a fifth more than the actual dip. The other deduction from the gross power applied for friction of the gearing between the motor and the wheel for the gudgeons and for water lifted too high they consider also as much overrated. Mr. Korevaer, in a paper read before the Dutch Institution of Civil Engineers, gives the useful effect of four scoop wheels and engines in the Netherlands at a mean of 67 per cent. of the indicated horse-power, the greatest being 69·6, and the least 60·0, the lifts varying from 3·66 feet to 6·0 feet. At Katwijk the percentage varied from 33 to 70 according to the height of lift, the mean being 50. At Gouda, with curved scoops, the percentage was 56·3 per cent.; with the wheels as recently altered to flat scoops, and with new engines an efficiency of 81·97 per cent. was obtained, the lift being 5·80 feet, and the quantity discharged 598 tons per minute. Mr. Huet, in his description of the scoop wheel at Adria, in Italy,* which has a diameter of 39½ feet and width of 6½ feet, with a lift of from 4 feet to 6 feet, states as the result of working that the proportion of horse-power of water lifted to the indicated horse-power is 72 per cent. In the Wexford Harbour trials, when the wheel was first started, the percentage of useful effect was 68·2.

* 'Stoombemaling van Polders en Boezems,' door A. Huet, C.E. Published at The Hague, 1885. The author here acknowledges much valuable information that he has obtained from this work.

CHAPTER VI.

THE ARCHIMEDIAN SCREW PUMP.

THESE pumps, although frequently used for lifting water from drains for emptying and cleaning them out, and other similar yurposes, have seldom been applied in this country for the permanent drainage of land. They derive their name from Archimedes, the Syracusan, who lived 287 B.C., and invented this machine, during his stay in Egypt, for draining and irrigating land. They were subsequently used by the Romans. The Dutch have used them very extensively in Holland for raising water for the drainage of the Polders.

The screw pump consists of three parts. A solid cylinder in the centre, called the core, to which is attached one or more spiral screws, and sometimes an external case. The number of screws running round the core varies from one in the simplest machine, to three or four in those of larger character. The ends of the core terminate in gudgeons which revolve in bearings, the lower one fixed under the water, and the upper on a beam spanning the delivery opening. As the efficiency of this machine is not affected by the speed at which it runs it is suitable for being driven by steam, wind, or hand power. In small pumps a crank handle is attached to the upper part of the core, and on this a pole with an eye through the centre, bushed with metal, is attached, the pole having cross handles at each end. One man works at the handle on the core, and one or more at each of the handles on the pole. It is reckoned that one man can raise in an hour at the rate of 1738 cubic feet of water 1 foot high, the pump making forty revolutions a minute. If worked by machinery, the pump is

The Archimedian Screw Pump.

driven by a spur wheel at the top geared into a bevel wheel and shaft.

The water level on the inlet side may vary without affecting the efficiency of the pump, except so far as the increased weight is concerned, due to the greater length required to meet the variation. But any change in the level on the delivery side immediately affects the efficiency. These pumps are not therefore adapted for use where there is much change in the level of the water into which they discharge.

The angle which the pump forms with the horizon when fixed varies according to the ideas of different constructors, but generally it may be taken that the most efficient position for the pump is when the angle of tilt is rather less than the spiral angle. Thus, for a machine having a spiral angle of 40° the angle of tilt for the pump should be 30°. The spiral angle is the form which the screw assumes with reference to the core, and is the angle made by a tangent drawn to the spiral on the cylindrical core, and a vertical line parallel to the axis of the cylinder. This angle varies from 30° to 60°. The Romans usually made it 45°. In the most effective machines it varies between 30° and 40°.

The discharging power of these pumps varies so much with the different circumstances under which they are worked, depending on the number of threads, the angle at which they are placed, the angle at which the pump works, and other matters, that it is difficult to give any precise formula for the quantity discharged. Upon pumps working under nearly similar conditions the discharge is as the cube of the diameter, and approximately it may be taken that, under favourable conditions, a pump 1 foot in diameter will discharge 0·32 cubic feet of water for every revolution. The number of revolutions varies according to the kind of power applied and the size of the pump. Small pumps of about 1 foot in diameter may be run at sixty revolutions a minute, the larger not reaching more than twenty. For drainage purposes it may be taken that these pumps can be run at from twenty to forty revolutions a minute. They have been used in Holland to lift the water 15 feet. Mr. Korevaer, a Dutch engineer, who has investigated the matter, places the limit of height at 14 feet, and the limit of discharge at 3500 ($98\frac{1}{2}$ tons) cubic feet per minute. The ten screws erected at Katatbeh, in Egypt, discharged 137·5 tons a minute each, making five to six revolutions a minute. The screws were enclosed in iron cases, but were found unequal to the weight of water they had to carry, and were consequently removed. Where the amount of water to be lifted exceeds the capacity of one pump it is customary to couple two or more together, all worked by the same engine.

The useful effect of these pumps is about the same as scoop wheels, and varies according to construction, from 50 to 80 per cent. of the power applied.

The Dutch screw pumps are constructed to work without an external casing, the wheel revolving in a semi-cylindrical trough of masonry. The weight of the water is thus borne on the masonry, and the screw is relieved of the strain. An example of one of these pumps is given in Plate 4.3

Mr. W. Airy, in a paper contributed to the Transactions of the Institution of Civil Engineers in 1871 (vol. xxxii.), gives the results of experiments carried out by him to test the relative merits of screw pumps of different construction. The results he arrived at were:—(1) That the smaller the spiral angle, *i.e.* the quicker the spiral, the flatter must the machine be laid to the horizon to produce its best effect. (2) That in an equal number of revolutions the quicker spirals will lift much more water than the slower ones. (3) That there is a great difference in the effect of the machines according as one end or the other is upwards. The advantage is greatly in favour of the machine when placed so that the acute angle which the thread makes with the core is downwards. With regard to the number of threads, Mr. Airy is of opinion that every machine should have as many threads as the conditions of ordinary workmanship and convenience will allow. That for screws of any size, say 6 feet or 7 feet external diameter, the width of the chambers should not be less than 18 inches on the square, and the diameter of the core one-third nearly of the external diameter. These conditions allow four threads for a screw, whose spiral angle varies between 20° and 30°; three threads for those between 40° and 50°; and two threads for 60°. The threads of the screws of the pumps upon which he experimented, and which he considered was the proper form to use, were made developable, by which term he meant a curved surface that could be unwrapped, laid flat, and inside a plane. The surface of the spiral thread, as ordinarily used, lies at right angles to the surface of the core, and if laid out flat the external edges would have more surface than the inner. Screws developed from a flat plate hold more water than those having threads at right angles to the surface of the core, and are easier to construct. The effect of the internal frictional resistance with a pump 3 feet in diameter, 10 feet lift, and running at twenty revolutions a minute, he found to vary from $4\frac{1}{2}$ to 8 per cent. of the useful effect realised; the gudgeons being 4 inches diameter absorbed 12 per cent. of the power applied; and that the best machines give off a useful effect of 85 per cent. of the power applied. For the most economical machine he considered that the spiral angle should not be less than 30° nor greater than 40°; and that the limit of height at which these pumps can be worked advantageously is 10 feet.

CHAPTER VII.

CENTRIFUGAL PUMPS.

IN giving the following description of the centrifugal pump, the subject has been confined, as far as practicable, to pumps with low lifts adapted for drainage purposes.

Although the principle of the centrifugal pump was known more than one hundred years ago, and pumps of this description were made and used experimentally fifty years since, it was not until the Exhibition of 1851 that they were brought into prominent notice. At the British Association meeting held at Birmingham in 1849, Mr. J. G. Appold exhibited a model of a centrifugal pump. After that, by an exhaustive set of experiments, principally directed as to the best form for the fans, Mr. Appold gradually improved the discharging capacity, and was enabled to exhibit a pump at the Exhibition of 1851 which formed one of the chief features of interest, the public being astonished at the immense volume of water put in motion by a machine which appeared, both from its size and simplicity of form, to be quite inadequate to the result attained.

Practically the form of pump shown at the Exhibition is the pump of the present day, no material alteration having been since effected. Although the general principle on which the pump acts is simple, the determining of the proportion of the different parts, and of the effect of the shape of the fan, is extremely difficult, requiring complex calculations, the data for which are so scanty as to render them not to be relied on unless checked by actual experiment.

Centrifugal pumps were first brought into use for the

drainage of land in consequence of the successful trials at the Exhibition. The proprietor of Whittlesea Mere, a large tract of fen and morass, was so satisfied with the performance of this machine that he gave instructions to Messrs. Easton and Anderson, the exhibitors, for the erection of an Appold pump, calculated to discharge 15,000 gallons—sixty-seven tons—a minute to a height of 5 feet. The lift of this pump had subsequently to be increased from time to time as the land settled, an operation performed with so little difficulty as to prove the adaptability of the pump for this purpose. A full description of this pumping station will be hereafter given.

The centrifugal pump is a machine consisting of an outer case having inlet and outlet pipes, in which revolves a fan at a high velocity. This high velocity adapts them well for gearing direct to engines running at high speeds. A very large displacement of water is effected in a short time. The machines are compact and occupy small space. The weight also being about one-twentieth that of a scoop wheel, the area of buildings required is small, and the cost of foundations is very inexpensive compared to those required for wheels. The first outlay is also considerably less. The average difference of cost of the pumping stations erected in Holland during recent years is 20*l*. per actual horse-power in favour of the pumps. Another great advantage of the centrifugal pump is that it readily adapts itself to the varying lift which must be encountered in most drainage stations, and automatically adjusts the work thrown on the engine as the lift varies. At first starting the engine drain is full, and at its highest level. The lift therefore being smaller, the pump discharges a larger volume of water; as the water in the drain lowers the lift increases and the quantity pumped diminishes in proportion, giving more time for the water to flow from the distant drains down to the engine drain and keep it fed. If pumping into a tidal stream the same effect takes place; as the lift increases the pump adjusts itself to the altered circumstances by sending out a less quantity, gradually regaining its original discharge as the tide falls. Further, when permanent settlement of the land occurs, the cost of adapting a pump is trifling; all that is necessary being the lengthening of the inlet pipe. Where proper precautions are taken no practical difficulty has been found to arise from weeds and other substances which find their way into the pump well. Pumps have now been running for the last thirty-five years, and performed their work efficiently and without trouble. At the Lade Bank station a manure fork is shown which, having been accidentally dropped into the feeder on the inner side of the grating, safely passed through the pump and came out without stopping the machinery or receiving any damage itself. At the pumping station, at Codigoro, North Italy, the body of a lad passed through one of the pumps. The boy had been missed from his home, and it was ascertained that he was last seen in the Valli near the main canal which leads to the suction wells of pumps. Some days afterwards, when one of the engineers had taken off the man-hole cover of condenser belonging to one of the pumps, he discovered the missing body, which had passed through the pump, but its further progress was impeded by the smallness of the condenser tubes. To prevent such accidents the approach to the pump should always be well protected by gratings placed across the entrance to the raceway.

There are two types of centrifugal pumps—the one similar to the Appold pump shown in the 1851 Exhibition having a horizontal spindle, and almost invariably fixed above the level of the water; the other of the turbine form, having a vertical spindle, the fan and case being submerged. All the larger stations in England have been fitted with the turbine form, but in Holland and Italy the pump with horizontal spindle has been most generally used. Of the three largest makers of drainage pumps in this country, Messrs. Easton and Anderson have adopted the turbine form for all large works, whereas both Messrs. Gwynne and Co. and Messrs. J. and H. Gwynne have used the other form.

The following description of a centrifugal pump is taken from a paper read by Messrs. J. and H. Gwynne at the British Association meeting, held at Norwich in 1868, and partly from the patent specification of the pump patented by them in 1868. Fig. 3, Plate 4, shows front and side sectional elevation of this pump. The pump consists of an outer case E, with a disc or impeller A, having six arms or blades B cast on a centre boss. A centre plate C springs from the boss and gradually decreases in thickness to a knife edge, bringing the separate currents of water into each side of the disc without producing an eddy or reflux. The arms are radial for two-thirds of the length, and curve off towards the periphery in an opposite direction to the line of rotation, in order to direct the water into the sweep of the case and prevent it rushing against the outer side of the discharge passages. Two rings D—see also Fig. 4—one at each side of the arms, form the bearing surface. The suction passages F F branch off from the suction pipe G at the point g. The bottom part of the casing E is thinned off to a knife edge, as shown at g, in order to prevent any obstruction to the water. A space is left between the passage and the case to carry the suction pipes F F over the enlargement of the discharge passage in a straight line to the openings in the centre of the disc A, at which point they curve into the top of the openings. The discharge passage is sprung from the periphery of the disc in the form of a helix or volute, commencing at the top of the case E and gradually increasing till it reaches the full size of the discharge pipe E^1. That part of the pump casing E which contains the impeller A is made of the same shape as the profile of the impeller, and similar in section and of just sufficient size to permit the impeller to revolve, the bearing rings D being accurately fitted to the turned recesses in the casing. By this means the usual side plates on the discs of centrifugal pumps are not required, the peculiar form of the pump casing acting in the place of such plates, consequently the friction of the disc A is reduced. The whole of the disc and arms are steel in one casting. The spindle passes through two stuffing-boxes cast on the casing E, to which are fitted gun-metal glands. A driving pulley is attached to the end of the spindle. Valves are placed at the bottom of the suction pipe to retain water when the pump is not at work. The area of the disc is equal to the area of the inlet and outlet pipes at all points. As these pumps, when working at their best speed, discharge every revolution three times the cubical contents of the disc, the discharge passages are three times the cubical capacity of the disc. By a simple arrangement the discharge pipe from the pump is made to act as a condenser for the steam from the engines. These pumps have been further improved in detail since the publication of the above description.

The points essential to a good pump, and common to all makers, are as follows :—The shape of the vanes to be such as to facilitate the movement of the water so that it shall enter and leave without shock, and shall therefore correspond as nearly as practicable with the path described by any fillet of it from the interior to the exterior of the fan; all changes of direction in the connections with the pump to be as gradual as possible, and all enlargements or contractions in the passages avoided; the passages throughout to be proportioned so as to have a gradually increasing velocity in the water until it arrives at the circumference of the fan, and then to have a gradually decreasing velocity until it issues from the discharge pipe; ample space to be allowed in the case, the larger the opening in the case the better the water passing off.

The chief attempts of makers to improve this form of pump have been in the form of the blades. All agree that these must be curved. The form of the curve varies with the ideas of the different makers. At the trial at the Exhibition of

1851 the Appold pump with curved arms of the form shown in Fig. 5, Plate 4, gave a maximum duty of 0·68, while the same pump with the arms made straight and inclined at an angle of 45 degrees to the radius gave a duty of only 0·463, and when the vanes were made straight and radial this duty was further reduced to 0·243. This pump had a 12-inch fan with internal width of 3 inches in two divisions of 1½ inch each. The central opening for the admission of the water on each side of the fan was 6 inches in diameter. The quantity of water discharged was 2100 gallons (9·37 tons) per minute at a height of 8·20 feet, and 681 gallons (3·04 tons) at a height of 27·60 feet, the fan making 828 revolutions in the former case, and 876 in the latter.

The pumps having horizontal spindles, from the ease with which they can be fixed, are eminently adapted for temporary purposes or for drainage in places where it is not considered desirable to cut through a river bank. For small drainage areas pumps of this type are preferable to those which are submerged, owing to the greater facility of access to the fan should it become choked with weeds, ropes, or other matter liable to be twisted round the blades. Substances escaping through the protecting gratings, which will readily pass through the openings in large pumps, frequently get wrapped round the fans and spindles of the smaller pumps; and if not actually bringing the pump to a standstill, seriously affect the efficiency and throw a greater strain on the engine. The action of a small centrifugal pump has been known to be completely stopped by eels, which had become twisted round the spindle. To remove these impediments without taking the case to pieces a hand-hole is frequently left in the cover, and so fixed on with screws as to be readily removable.

Pumps should be so fixed that all the parts can be readily accessible, and the suction pipe provided with means for slinging it with a rope or chain when required to detach it from the pump. The side plates are put together with bolts

H

and nuts, and in case of any substance finding its way into the pump sufficiently large to cause a stoppage, the taking to pieces and putting together is a work occupying only a very short time. The bearing of the driving shaft should be taken off the pump case by a sliding iron seat fixed close up to the case, which can be pushed away when the glands of the pump require packing.

Pumps of this description require charging with water before they can be started. This can be done by a small donkey pump, or by exhausting the air by a steam jet. In the latter case the outlet pipe and flap or non-return valve, which in all cases is necessary to prevent the back-flow of the water when pumping is stopped, are made sufficiently air-tight to allow the water to flow up and fill the pump. In the former case a valve opening inwards is placed on the inlet, which has also a perforated rose at its termination to prevent, as far as possible, the entrance of foreign substances. In large drainage pumps a special air pump is provided for charging the pump.

In order to avoid unnecessary lifting of the water the discharge pipe is carried down below the lowest level of the water on the outer side. The lift is then not greater than the difference of level of the inner and outer channels, the suction and delivery pipe acting as a syphon.

The turbine form of pump has a vertical spindle, and must be placed below the water, at the lowest level from which the water has to be pumped. The earlier forms were made with the fan divided into two parts with a centre diaphragm, but pumps of more recent construction have only a single fan with one inlet. When the water enters on one side of the wheel only, it causes a thrust in that direction. This thrust is made use of to counterbalance the weight of the driving shaft and pump vanes; then after the pump is once in motion all weight being thus taken off the bearings. For the same quantity of water to be discharged the single pumps necessitate the fans being made of larger diameter than those having double fans.

Centrifugal Pumps.   99

Pumps with single fans can therefore be run at a smaller number of revolutions—a great advantage when used for drainage purposes. The water passes to the opening through a trumpet-shaped mouth continued downwards a short distance to prevent the pump drawing air when the water is pumped to the lowest level. The gradually decreasing size of the opening at the entrance produces a corresponding increase in the velocity, which is again decreased on leaving the fan by guide blades, the apertures between the guide blades being smaller near the fan than above.

The use of guide blades is found to increase very materially the discharging power of the pump. In experiments made by Mr. Parsons, a pump, which without the guide blades discharged 1500 gallons a minute, increased this quantity to 5000 gallons a minute when the guides were added, the pressure of steam remaining the same.

This form of pump is fixed in an iron case or brick well, the outlet from which is at the lowest level to which the water in the outlet channel is ever likely to fall. No delivery or suction pipes are required. The opening in the well is either protected by a self-acting valve to prevent back flow when the pump is not working, or doors are placed at the end of the channel leading into the main outfall drain. The pump is hung by its spindle to a girder across the well at the top by a gun-metal bearing that is quite accessible, the spindle being stayed by horizontal guides in the well. No footstep is required. The bearings of the different parts have conical seats, and the fan can at any time be taken out for repairs and replaced without emptying the water from the pump well. This arrangement secures the machine from the wear and tear due to its exposure to the grit, and dirt contained in the water, and facilitates repairs when required. It is necessary to arrange the well or case so that it can at any time be pumped dry if required—a precaution, however, that is seldom wanted. No valves are required, the pump being always charged and ready for starting; being also covered by a considerable depth of water it is free from the action of frost, which is liable to freeze the water if left in a pump exposed to its action, and burst the case. The friction of the water along the suction and delivery pipes necessary in the other form of pump is also avoided. On the other hand, the gain in efficiency due to a properly formed pump case is sacrificed.

The action of a centrifugal pump is as follows:—As soon as the fan begins to revolve the blades carry the water with them, which is then pushed forward and drawn into the case partly by the mechanical action of the blades propelling the water forward, and partly by the centrifugal action created by the rapid rotary motion created by the fan. The vacuum created by the water driven out of the fan is immediately filled by a fresh supply of water from the inlets. The water driven out by the fan is propelled along the discharge passage, and having no other means of escape, rises up the pump well till the outlet is reached. A constant and continuous stream without check or shock, as in bucket pumps, is thus created, and the water is kept in motion along its whole passage.

The best velocity for the water to flow through the passages of the fan is from 6 feet to 8 feet per second. The discharge increases with the increase of velocity, a small increase in the number of revolutions producing a large increase in the delivery. Mr. Parsons* states that he found an increase of 14 revolutions—392 to 406, or about $3\frac{1}{2}$ per cent.—increased the discharge from 1012 gallons to 1753, or 42 per cent. Up to a certain speed the pump does not act, and the fan revolves without lifting the water over the overflow. Unless the speed for which the pump is intended to be run at is attained, the machine does not work at its best, and fuel is wasted. It is important therefore that the engineman in charge should know the velocity for which his pump is

* 'Trans.' Inst. C.E., vol. xlvii.

speeded. The formulas for ascertaining the quantity of water discharged by a centrifugal pump are complicated, and the coefficients vary with each particular make. Messrs. Easton and Anderson furnish the purchasers of their drainage pumps with a diagram by which the quantity discharged can be ascertained by inspection when the lift and speed are known. If an account is kept of the lift, and the speed and time of working of the engines by the engine-driver, checked by locked counters attached to the machinery giving the number of revolutions, a means is provided of preserving a record of the quantity of water pumped, and at the same time a check is placed on the engine-driver.

In the centrifugal, as in all other pumps, a certain amount of the power applied in driving them is absorbed by friction of the bearings and resistance due to roughness of the surface of the pump, and to the slip and eddying motion of the water; this loss varies from 30 to 50 per cent. The duty to be expected from a centrifugal pump of the best construction used for drainage purposes may be taken at about 70 per cent., falling in small-sized pumps, and those not of the best construction to 50 per cent. The power absorbed by an engine and pump may be divided approximately as follows:—

Friction of engines	10 per cent.
„ pumps	30 „
Efficiency	60 „
	100

The ratio of useful effect in water lifted, as compared with the indicated horse-power, for the direct-acting engines and pumps put up by Messrs. J. and H. Gwynne, has been found to vary from 55 to 70 per cent.

The efficiency of a centrifugal pump rapidly diminishes if the lift is greatly increased. From observations made by Mr. Richards of the pumps used for lifting water for irrigating purposes on the Pacific coast, he obtained an efficiency of from 65 to 70 per cent. for lifts up to 10 feet, and only 35 per cent. for lifts of 100 feet.

The statical height of water which these pumps will support without discharging requires a speed which varies in some degree with the form of the blades. Mr. Thompson* states, as the result of his experience, that the speed of the periphery per second required to balance the weight of the water up to the point of discharge is equal to eight times the square root of the given height in small pumps, and 9·82 times in large pumps. In a letter which appeared in 'The Engineer' of September 24th, 1886, Mr. C. Brown gives, as the result of experiments made by him with pumps having blades of the form shown in Fig. 6, Plate 6—the water being held at a height of 45 feet—

(1) Required a speed per second $= \sqrt{2gh}$.
(2)   ,, $=$ considerably more.
(3)   ,,   still more.
(4)   ,, $0·82 \times \sqrt{2gh}$.

In a description of the pump made by Mr. C. Hett, of Brigg, for the s.s. Eldorado, in 'The Engineer' of June 18th, 1886, it is stated that Mr. Hett found a pump of his make with 2 feet disc which gave a full discharge at a height of 16 feet 6 inches when running 190 revolutions a minute, the velocity at the periphery being only about two-thirds of the head due to gravity.

The following formula of Unwin's for ascertaining the speed and discharge is given in Molesworth's 'Pocket Book':—

$S = 8 \sqrt{H}$ in small fans, or $9·5 \sqrt{H}$ in large fans.

$H = \dfrac{S^2}{64}$,,   or $\dfrac{S^2}{90·25}$,,   ,,

---

* 'Trans.' Inst. C.E., vol. xxxii.

$$D = C\sqrt{\frac{Q}{\sqrt{H}}}.$$

D = Diameter of fan in feet.
H = Head of water in feet, including head, corresponding with friction of pipes, &c.
S = Speed of periphery of fan in feet per second.
Q = Quantity of water lifted, feet per minute.
C = ·12 to ·18.

The great advantage that a centrifugal pump has over all other machines for raising water for the drainage of land where the lift is constantly varying, either from the rise and fall of the tide in the outfall river or the lowering of the water in the inside drains as the pumping proceeds, is that it adapts itself to these variations in the lift without any alteration in the speed of the engine, employment of differential gear, or attention on the part of the engine-driver. If kept working at the ordinary speed, the pump will discharge either more or less water as the lift diminishes or increases.

Centrifugal pumps of the smaller class are generally kept in motion by a strap running round a pulley on the spindle of the pump and the driving wheel of the engine. Those of the turbine form are worked by a mitre pinion, keyed on to the vertical spindle gearing into a bevel wheel on the crank shaft of the engine. Direct-acting engines and centrifugal pumps are also constructed with engine and pump on the same base plate, the piston-rod being attached by a short connecting-rod to a crank in the spindle of the pump. On Fig. 7, Plate 4, is shown one of a pair of these pumps as fitted up by Messrs. J. and H. Gwynne for the drainage of the Grootslag Polder, near Andyk, the lift being 10 feet 6 inches and the discharge 75 tons per minute for each pump.

Messrs. Gwynne strongly advocate the use of the pump with horizontal spindle as preferable to the turbine type, being, in their opinion, more effective, occupying only a very small space, and requiring inexpensive foundations, the cost of the instalment being considerably less than that of any other type.*

Messrs. J. and H. Gwynne have erected several of these direct-acting engines and pumps in Holland and France, particulars of some of which are given in Chapter VIII. For the drainage of the Middel Polder in Holland, containing 1600 acres, a pair of their engines and pumps were erected in 1878. The space occupied by engines and pumps is only 21 feet by 11 feet, and they are each capable of delivering 30 tons a minute to a height of 16 feet. Although these engines make 138 strokes a minute, and, when emptying the lake, worked for three months, night and day, they have run for eight years without anything being required to be done to them except ordinary repairs. At the Legmeer Polder a pair of direct-acting vertical engines and pumps were also erected by the same firm in 1875, of which an illustration is here given, Plate 4. Each engine and pump is capable of lifting 75 tons 17·38 feet per minute, and occupies a space 15 feet by 10 feet. These engines make 156 revolutions a minute; yet, notwithstanding this high speed, the author was assured by the superintendent in charge that, beyond an accident to the fan of one of the pumps, owing to a piece of wood having got into the case, no stoppage had taken place, and the repairs had only been slight and of the ordinary character common to all machinery. The satisfactory working of this class of pump was confirmed at other stations in Holland which the author visited. The small space occupied by this class of machinery, and the fact of the pump being placed on the same floor with the engine, is the means of effecting considerable saving in foundations. In the case of the pumps at Middel Polder, a road intervenes between the engine-house and the river into which the drainage water is discharged; the iron delivery pipes are carried beneath this road, the

* 'Notes on Pumping Machinery for Drainage Purposes,' by J. and H. Gwynne, 1885.

Centrifugal Pumps.

usual masonry culvert under the road being thus dispensed with.

The theory relating to the discharge of these pumps, with the result of experiments in connection therewith, will be found in the papers in the 'Transactions' of the Institution of Civil Engineers, by Mr. Thompson, in 1871, vol. xxxii.; Mr. Parsons, in 1876, vol. xlvii.; Mr. Unwin, in 1877, vol. liii.

The various sizes of these pumps are generally described from the diameter of the inlet pipes. Thus a "30 inch pump" would be one having a suction pipe 30 inches in diameter at the inlet to the pump.

With the best class of pumps with low lifts, such as are required for drainage purposes, the following may be taken approximately as the rate of discharge. The amounts given are, however, above those attained by the pumps in ordinary use :—

Diameter of Suction and Discharge Pipe, in Inches.	Water discharged per Minute.	
	Gallons.	Tons.
15	5,000	22·32
18	7,000	31·25
24	11,000	49·10
30	18,000	80·35
36	20,000	89·28
42	27,000	120·53
48	40,000	178·57
54	70,000	312·50
60	100,000	446·43

CHAPTER VIII.

PUMPING STATIONS.

THE pumping stations described in the following chapter have been selected as fairly representing the various types of machinery in use for draining low land, and for raising water for irrigating purposes.

It has been difficult to obtain reliable average results as to working, owing, in many cases, to the want of accurate records of working, and also to the fact that the work done is constantly varying from the daily rise and fall of the water. A further difficulty occurs from the different ways of expressing weights and terms in this and other countries. The author has endeavoured, as far as practicable, to arrive at correct results. In order to facilitate comparison, the data are reduced in every case to one common standard of coal consumed per horse-power of water lifted and discharged per hour, and of tons of water lifted a given height in feet per minute. A standard is thus afforded for readily comparing the different kinds of machinery in use, and also showing whether a pumping station is being worked with a due regard to efficiency and economy.

PODEHOLE, DEEPING FEN, LINCOLNSHIRE.—The taxable area of this drainage district is 30,000 acres, the quantity of land actually draining by the wheels being 32,000 acres. The water from the fen is collected into two large drains, from which it is pumped into an outfall cut, called the Vernatt's drain, which discharges into the tidal river Welland, about six and a-half miles distant. The machinery was erected in 1824, and consisted of two scoop wheels worked by two low-pressure condensing beam engines of 80 and 60 nominal horse-power respectively, working at a maximum pressure of steam in the boiler of 4 lb. This pressure has since been altered and other improvements made. The crank shaft from the engine passes through the wall of the engine-house, and carries a pinion gearing into a spur wheel on the shaft of the scoop wheels. The ratio of the velocity of the engines to the wheels is 16 to 5, and 22 to $4\frac{1}{2}$ respectively. The larger engine—called the Holland—has a steam-jacketed cylinder, 44 inches in diameter, with 8 feet stroke. The fly-wheel is 24 feet in diameter. The smaller engine—called the Kesteven—has a steam-jacketed cylinder 45 inches in diameter and 6 feet 6 inches stroke. The fly-wheel is 24 feet in diameter, making 22 revolutions a minute. The framing of the scoop wheels is of cast iron. The larger wheel was originally 28 feet in diameter, and fitted with forty scoops, but the diameter was increased about ten years ago to 31 feet. The scoops are 6 feet 6 inches long—radially—by 5 feet wide, giving an area when wholly immersed of 32·5 square feet. The mean diameter is 24 feet 6 inches, number of revolutions a minute 5, giving a gross discharge, after deducting the space occupied by the scoops, of 11,215 cubic feet per minute, or 313 tons. These wheels, as running at the present time, have been very accurately fitted in their places, and run very true, so that there is a clearance of barely half an inch between the floats and the masonry at the bottom and sides. The smaller wheel is 31 feet diameter, with the same number of scoops, each being 5 feet 6 inches long by 5 feet wide, giving an area of 27·5 square feet. The mean diameter is 25 feet 6 inches, number of revolutions a minute $4\frac{1}{2}$, equal to a discharge, after deducting scoops, &c., 8959 cubic feet per minute, or 250 tons. The scoops dip from the radial line at an angle of 25°, being tangents to a circle 7 feet 6 inches, in diameter. This angle being found too small to give the best results, the end of each scoop for a length of 18 inches was altered so as to dip further back 6 inches. The straight part of the scoops enters the water at average flood level at an angle of 29°, and leaves it at 36°. The average dip in floods is 5 feet, and the average head 5 feet, rising to 7 feet in extreme floods. Steam is supplied to the engines by five double-flued Lancashire boilers, having water-pockets above the furnaces; they are 7 feet diameter by 26 feet long. The total discharge of the two wheels is 563 tons per minute. This is equal to about the fourth of an inch of rain over the whole area of 32,000 acres when the wheels are working to their full capacity for twenty-four hours a day.

The efficiency of these wheels has been greatly increased by alterations carried out a few years ago. On the inlet side—see Plate 5 —a shuttle has been added, by which the amount of water coming to the wheel can be adjusted and the supply regulated to the quantity best adapted for keeping the wheel fully charged without its being drowned by it. This shuttle is of the same width as the wheel, and consists of a wooden door fixed across the inlet close up to the wheel, and working on friction wheels in a frame placed in the masonry. The door is fixed close to the wheel, at an angle of 45 degrees to the bottom of the raceway. It is provided with a balance weight, hung by a chain working over a pulley. The shuttle is lifted or lowered by a toothed rack gearing into a spur wheel and pinion attached to a shaft, which is carried up into the inside of the building. The floor drops away from the bottom of the shuttle on the inlet side in a circular form, so as to give a larger space for the admission of the water, and allow it to come up and pass freely under the shuttle. The water passing under the shuttle does not catch the scoops until they come towards the bottom of the trough, and then impinges on them in the same direction in which they are travelling, and with a velocity due to the head of water at the back of the door, and thus aiding in the forward motion of the wheel. The scoops become fully charged as they assume a vertical position. The apparent increase in the lift from the lower level from which the water has to be raised is more than compensated for by the avoidance of the mass of dead water which a wheel generally has to encounter on first entering the water, and by the wheel being just sufficiently fed with water having a velocity and direction which assist in sending it round. A much greater quantity of water is thus raised with the same amount of steam than could be done if the shuttle were not there. With the surface of the water in the inlet drain during floods standing 6 feet 10 inches above the bottom of the scoops, the shuttle is lifted sufficiently to allow 1 foot 3 inches of water to pass under it, and this keeps the wheel well supplied. A movable breast has also been fixed on the outlet side. It is made of iron plates, and works into a recess cut in the masonry of the breast, so that its face is flush with it. The plates are bent so as to have the same radius as the wheel; the upper part of the segmental plate is hinged at the top into another flat wooden platform fixed to an iron frame, which when down lies in a recess in the floor of the outlet, and rises with the breast. To enable this platform to adjust itself to the space in which it has to lie, it is so formed that one end slides in and out of the iron frame.

Pumping Stations.

The lower end of the frame is hinged to the floor; thus, when the breast is raised the floor is also raised for some distance, forming an inclined plane from the top of the movable breast to the floor of the outlet channel. The breast is raised or lowered to adapt it to the height of the water in the outlet drain by a segmental toothed rack gearing into a spur wheel attached to a windlass fixed on the wall of the raceway. By raising this breast to a sufficient height to allow of a free egress of the water over it, the back current at the bottom of the outlet, which always exists with the old arrangement, is entirely avoided. These improvements to the wheel have been carried out under the direction of Mr. Alfred Harrison, the superintendent of the Deeping Fen drainage district.

During the five years, 1876–80, the average work of the two engines amounted to 219½ days of twenty-four hours each for one engine, and the consumption of coal averaged 5 tons 9 cwt. per day These engines have lately been thoroughly overhauled by Messrs Watt & Co., and new boilers provided, the working pressure of the steam being raised to 20 lb. on the inch. The coal consumption has been reduced to 3·28 tons per day, the amount of work done by the engines being at the same time very largely increased. It was reported that, owing to these improvements, 60 per cent. more water was raised with 42 per cent. less fuel. The annual saving was estimated at 450*l*. in wet seasons.

The average annual cost of this pumping station for the three years 1880–83, when the rainfall was considerably above the average, was 1412*l*., of which 1009*l*. was for coal, which cost about 15*s*. a ton. The average quantity consumed during the three years was 1356 tons per year. Taking the area drained as 32,000 acres, this gives 23·61 acres for each ton of coal. The cost per acre is 10·58*d*., or taking coal only, 7·56*d*. Taking the average lift at 5 feet, this gives 1·51*d*. per acre per foot of lift for coal only. The following is the time the engines worked during the above period :—

	80-H.P. Engine.	60-H.P. Engine.	Coals consumed.	Rainfall.
	hours	hours	tons	inches
1880–81	5112	3912	2104	37·12
1881–82	2616	1680	718	26·12
1882–83	2664	3756	1317	32·87

Taking the latter period as a fair sample of a wet season, and allowing the average dip of the wheels throughout the whole period the wheels were running to be 2 feet 6 inches, and the head 4 feet 6 inches, the average work done in water lifted would be 83·63 H.P. The average consumption of coal, 442 lb. per hour, equal to 5·28 lb. of coal per hour per horse-power of water lifted and discharged.

LADE BANK, LINCOLNSHIRE.—This pumping station is for the drainage of a district known as the East Fen, forming part of the system of the Witham Drainage Trust. It was drained and brought into cultivation at the beginning of the present century, the principal drain being fourteen miles in length, and discharging into Boston Haven through a sluice with three openings of 15 feet each, the outlet doors being self-acting. Owing to the subsidence of the peat in the fen, the drainage of this district became imperfect, and in wet seasons it was frequently flooded, the proprietors in several cases using scoop wheels driven by portable engines to lift the water off their land, the aggregate power of these engines amounting to 80 horse-power. It was consequently decided to provide pumping machinery for more effectually draining the lowest parts of the district. In 1867 the pumping-machinery was erected, the site being fixed at Lade Bank, the pumps discharging into the main drain about nine miles above the outfall sluice. The area of land which is pumped is 35,000 acres. The average lift is about 4 feet, the extreme being 5 feet; and it was assumed by Sir John Hawkshaw, under whose direction the works were carried out, that pumping power should be provided equivalent to lifting a continuous rainfall of a quarter of an inch in twenty-four hours over the whole district. The machinery consists of two pairs of high-pressure condensing vertical and direct-acting steam engines of 240 aggregate nominal horse-power. Two massive A frames span over either side of the pump well, and carry the crank shaft, on which is fitted a large mortice bevel fly-wheel. The cylinders, which are 30 inches diameter by 30 inches stroke, are placed outside either A frame, being carried on a heavy base plate. Two small A frames fixed on the cylinder covers carry the parallel motion of a wrought-iron grasshopper beam, one end of which is attached to the crossheads of the piston-rods, the other end being carried on a vibrating column.

From this beam the air-pump and feed-pump are worked. The side valves are worked by means of excentrics on the crank shaft, situate just inside the A frames. The arrangement of one of these engines is shown in the sketch, Plate 5. The bevel mortice fly-wheel gears directly into a pinion on the pump spindle, which is suspended from a bracket, spanning across the engines by means of an onion bolt bearing. By this arrangement, not only can the fan be readily withdrawn, but the bolt allows of any necessary adjustment in the level of the fan. Steam is supplied by six Lancashire boilers 23 feet by $6\frac{1}{2}$ feet, the furnaces being 5 feet long by $2\frac{1}{2}$ feet, the working pressure being 50 lb. to the inch, steam being cut off in the cylinder at quarter stroke. The base plates of the engine are partly supported by the brickwork, and rest on and are bolted to the cast-iron cylinder, which forms the lining of the pump-well. There is one pump well to each pair of engines. The pump case consists of a cast-iron cylinder, 12 feet in diameter, 9 feet 6 inches deep, open throughout its whole depth on the delivery side, and furnished with self-acting gates, 12 feet wide. In each well is a double-inlet Appold centrifugal pump. The fan is placed horizontally, and is 7 feet in diameter and 2 feet $4\frac{1}{2}$ inches wide, the mouth of the lower suction pipe being 3 feet 6 inches above the floor of the well, and 4 feet 6 inches below the surface of the water at the ordinary drainage level. The upper suction pipe curves over, the mouth being about 1 foot 6 inches above the other. Each pair of engines and pumps works independently, and is capable of lifting 350 tons of water a minute 4 feet 6 inches high, being the largest amount in volume for one pump which had been erected at the time. The engines are placed in a brick building 34 feet by 46 feet, and 18 feet high. The boiler house is 69 feet by 38 feet. The chimney shaft is square, 90 feet high, and 4 feet 9 inches inside at bottom. The foundations rest on a bed of Portland cement concrete. Across the main drain are two sluices, each 12 feet wide, having doors to shut against the water on the lower side, and a lock 70 feet long by 12 feet wide, for the barges which navigate the main drain. The surface area of the main drains between the pumping station and the outfall sluice is about 100 acres. The machinery, buildings, and lock were erected by Messrs. Eastons, Amos, and Anderson, and cost 17,000*l*.

Taking the work done as 700 tons lifted 4 feet 6 inches high per minute, this gives 80·37*l*. as the cost per horse-power of water lifted.

The following account of the working of these pumps, a few years after their erection, was given by Mr. E. Welsh, the engineer to the Commissioners ('Trans.' Inst. C.E., vol. xxxiii.) :—

	Years ending March 31st.	
	1871.	1872.
Weight of water discharged in tons	13,564,190	18,2,6,130
Average lift in inches	44·77	45·00
Average revolutions made by engines per minute	36·02	38·20
Sum of hours worked by both pumps	794·25	980·5
Coals consumed during working hours in tons	328·00	397·25
Engine oil used, gallons	25·75	20·25
Tallow used, lbs.	181·	135·
Waste used, lbs.	135·	85·
Wages paid first and second drivers yearly	£158 12 0	£158 12 0
Boy, yearly	15 12 0	18 14 0
Fireman, 2085½ hours at 3½d., and 2033 at 3½d.	0 8 0	29 13 0

Taking the above account of work done and coal consumed, the horse-power of water lifted for both engines is equal to 72·52 horse-power for 1871, and 79·17 for 1872, the coals used equal to 11·37 lb. per horse-power of water lifted for the former year, and 11·46 lb. for the latter. This seems a very large consumption of coal for machinery of this class, but the correctness of the result is borne out by the quantity used by the engines and pumps for the North Sea Canal, in Holland, which are similar to these, and which are reported as using 11 lb. per horse-power of water lifted.

In 1875 there occurred a heavy flood in this district, the total quantity of rain registered for October and November was 9·49 inches. To cope with this, both pumps were running continuously from November 14th to the 20th, after which one pump only was used. The two pumps were running 177 hours, and one pump for 562, during which time 300 tons of coal were used.

In the flood of 1876–77 the engines were running from December 27th to January 11th; the highest lift being 5 feet 2 inches, the lowest 3 feet 3 inches, and the average during that period 4·20 feet.

For the three years ending 1881–83 the average working charges were 1089*l*., equal to 7·46*d*. per acre, taking the average lift at 4 feet, equal to 1·86*d*. per acre per foot of lift.

LITTLEPORT AND DOWNHAM.—This district consists of 28,000 acres of peaty fen land, situated in the South Level of the Bedford Level in the county of Cambridge. In addition to the fen land, some of the adjacent higher land also discharges its water into the drains

Pumping Stations. 113

of this district, so that the total area drained is about 35,000 acres. There are fifty-three miles of main drains which collect and convey the water to the engines, and twenty-three miles of catchwater drains. The drain which leads direct to the engine has a 22 feet bottom with slopes of $1\frac{1}{2}$ to 1. There are two pumping stations seven miles apart; one on the Hundred Foot River and one on the Ten Mile River—part of the Ouse—near Hilgay. The drains communicate with both, so that the water can flow to either station. The machinery at both stations consists of scoop wheels driven by beam engines, and was put up by the Butterley Company in 1830, under the direction of Mr. Glynn.

The Hundred Foot Station.—The scoop wheel at this station pumps into a tidal stream, and is now the largest in diameter that the author knows of. The wheel as originally constructed was 37 feet in diameter. It was subsequently altered to 41 feet 8 inches diameter, with scoops 2 feet 8 inches wide. This was removed and replaced by the present wheel, which has sixty scoops and a diameter of 50 feet, with an internal spur wheel of 36 feet diameter, gearing into a pinion on the crank shaft. The scoops at the same time were widened, and the radial length increased to 6 feet 6 inches and width of 3 feet 4 inches. The start posts are of oak, 7 inches by 4 inches, diminishing at the outer edge to 4 inches by 4 inches. The average dip of the scoops is 3 feet 3 inches; the greatest, 5 feet 6 inches; the average head, 13 feet 9 inches, the maximum being 17 feet. The scoops dip from the radial line at an angle of 42°, being tangents to a circle 25 feet diameter. At average flood level they enter the water at an angle of 31°, and leave it at 50°. With the maximum dip of 5 feet 6 inches, they enter at 22°, and at the maximum lift of 17 feet leave the water at 42°. The wheels make three revolutions a minute to 21 of the engine. This wheel has been very accurately hung on its bearings, the clearance on the delivery side between the wheel and the masonry at the sides and bottom being only about $\frac{1}{2}$ inch. On the inlet side the walls diverge from the wheel, the idea being to allow the inflowing water to get freely to the wheel to feed it. It is questionable whether the effect of this, by diminishing the velocity with which the water approaches the wheel, does not do harm. When working to its full extent the wheel discharges 197 tons per minute. A movable breast struck to the radius of the wheel, with curved top, worked by a shaft and gearing from the engine-house has been added to the original structure on the delivery

I

side of the wheel. This can be raised at pleasure 8 feet above the masonry delivery sill, which is 10 feet above the tips of the scoops. No portion of the floor is raised with the breast as at Podehole. This breast is so raised and lowered that its crest shall be below the level of the water in the outfall channel a depth equal to one-half the dip of the wheel, this proportion diminishing as it approaches towards the full height to which it can be raised. At a trial made in 1872 by Mr. Mason Cooke, superintendent of the district, it was proved that the use of this movable breast added most materially to the efficiency of the wheel. A temporary weir was fixed across the outlet channel at a sufficient distance to allow the water to get well away from the wheel. The crest of this weir was 7 feet 8 inches above the masonry delivery sill, or 17 feet 8 inches above the points of the scoops, equal to a high-flood level in the river. Steam during the trial was kept at a uniform pressure of 5 lb. in the boilers. The dip of the scoops was 3 feet 5 inches, and head 15 feet 11 inches. With the movable breast down the engine was not able to raise the water over the dam, but came to a standstill. The movable breast was then raised 4 feet, when the engine made 10 revolutions a minute; when raised to 5 feet the number increased to 12; at 6 feet, to 13; at 7 feet, to $13\frac{1}{2}$; and at 8 feet, or 18 feet above the tips of the scoops, the engine made nearly 14 revolutions per minute, and discharged over the dam a stream of water 7 feet 6 inches wide by 1 foot 8 inches deep. A vertical door is fixed on the inlet, which can be raised or lowered by gearing, but this is not used to regulate the inflow of the water, and the wheel is not provided with any adjustable shuttle. The wheel takes up and discharges its water quietly and efficiently, and although necessarily a portion of the water is lifted above the level in the outlet channel, there is no dashing about of the water, but it leaves the scoops freely and in a solid mass. The wheel is driven by a condensing beam engine, having cylinders 3 feet $7\frac{1}{2}$ inches diameter and 8 feet stroke; the steam pressure in the boiler is 20 lb., the pressure before the recent alterations being 5 lb. This engine was overhauled and refitted by Messrs. Watt & Co., of Birmingham, in 1881, and adapted to the increased size of wheel. The old cast-iron crank shaft was replaced with a wrought-iron crank with 4 feet throw, a circular slide valve to work from the parallel motion was substituted for the D valve, and an internal expansion valve added to cut-off at from one-tenth to one-half of the stroke, as regulated by a hand-wheel and screw. When pumping at an average

Pumping Stations.

level, the engine is well above its work, and advantage is taken of the expansion gear, the steam being frequently cut-off at from one-sixteenth to one-tenth of the stroke. Steam is generated in three Lancashire boilers, 24 feet long by 7 feet diameter, the safety-valves of which are adjusted to blow off at 20 lb. The enormous weight of this machinery compared to the work done may be judged from the following:—The wheel alone weighs 75 tons; the beam of the engine is 3 feet 8 inches deep, and weighs 15 tons; the fly-wheel is 25 feet diameter and weighs 30 tons.

The improvements in the engine and wheel resulted in a very considerable saving of coal, the consumption for 1881, before the improvements were made, being 1411 tons for a running of 2988 hours, and an average dip of the scoops of 2·66 feet as against 691 tons for 1883 for 2288 hours' running, and an average dip of 3·30, or, after allowing for the difference in the number of hours' run, this shows a clear saving of 369 tons of coals in one season, while the average extra head pumped against was increased 1·54 feet, or an increase of work of 31 per cent., and a decrease of coal of 35 per cent.

A series of trials was made by Messrs. Watt & Co., Birmingham, on January 13th, 1882, after the improvements to the engine and wheel had been completed by them. The following figures give the results of the last of these trials:—

Mean dip of scoop	4·0 feet.
Mean lift of scoop	14·79 feet.
Mean revolutions of engine per minute	18·66.
Mean revolutions of wheel per minute	2·75.
Mean velocity of water flowing down engine drain per minute	45·766 feet.
Area	137·397 feet.
Quantity per minute	6288·11 cub. feet.
Horse-power, gross indicated	224·05.
Horse-power, water lifted	176·158.
Ratio $\dfrac{\text{W.H.P.}}{\text{I.H.P.}} = \cdot 7862.$	
Mean circumference of immersed portion of wheel	144·5 feet.
Cubic contents of immersed portion, each revolution less space occupied by scoops	1782·11 cub. feet.
Ditto ditto per minute	4900·80 cub. feet.

The diagram, Fig. 9A, Plate 6, is the mean of those taken for this trial.

From this it will be seen that the quantity delivered, as measured by the water passing down the drain, instead of being less than that due to the theoretical discharge as measured by the wheel, was about 21 per cent. greater. Messrs. Watt observe as regards this:— "We took every care possible in getting at the true velocity. The drain—mean width 28 feet 9 inches by 4 feet 9 inches deep—was new, and entirely cleared out and free from weeds or obstructions. We put three floats, which were regulated by tubes to sink from the surface to within a few inches of the bottom, one in the centre, and one on each side; they just projected out of the water and had a feather put into the top so that there was the least possible obstruction from wind. All the floats practically went down the stream at the same pace, but if there was any difference the mean of the advance was taken. We have therefore every reasonable confidence that the number of cubic feet that passed down the drain is not far from being correct. As regards the allowance of 10 per cent. usually allowed for leakage, this is altogether in excess. The breadth of the ladles in the wheel is 40 inches, the clearance on each side is $\frac{1}{2}$ inch; the area therefore for leakage is only $2\frac{1}{2}$ per cent. upon the width of the ladle. Then the head which would cause any leakage is very small, being only that between the level of the water in one space between the ladles and the level in the next succeeding space; moreover, these ladles act something like a pocketed piston having grooves in it, and the general motion of the wheel, and the upward current of the whole mass reduces this leakage to a minimum. In a broad wheel it is practically imperceptible. As regards the dip, we found by very careful inspection that the ladles were more full by at least 1 foot than was represented by the mere dip; the fact being that the velocity imparted to the water by the outer diameter of the ladles, combined with the gradual contraction of the drain to a line passing through the centre of the wheel increases the velocity of the water, the result being that the cavities between the ladles are filled almost to the inner lining, so that the quantity lifted in the 100 Foot wheel would be represented by 5 feet depth in the ladles instead of 4 feet. At Podehole, where the feeding sluice or shuttle is depressed owing to the great depth of water in the feeding channel, it allows enough water to pass with the velocity due to the head, so as actually to fill the ladles quite full. The calculated discharge of the wheel, with 5 feet dip, and the other figures as given, would then work out at 6089·23 cubic feet, or equal to 76 per cent. of efficiency, as against 6288·11 cubic feet and efficiency of 78·62 per cent. as measured in the drain. The discrepancy between the power of the wheel to lift, and the quantity of water delivered in the feeding drain, are therefore reasonably consistent with each other." If the work be taken as that measured by the wheel with the dip of 4 feet, the efficiency would be only 61 per cent.

The Ten Mile Station.—The scoop wheel at this station is 43 feet 8 inches diameter, having been increased 20 inches from the original dimension by lengthening the scoops. There are 50 scoops, 7 feet 6 inches radial length by 3 feet wide. The average dip of the scoops is 3 feet; the greatest 5 feet 6 inches; and the lift 11 feet average, and 14 feet maximum. This wheel lifts the water into the Ten Mile River, which is not tidal, the tide being shut out by sluice doors at Denver Sluice. There is, however, a considerable rise in the river during tide time. These scoops dip from the radial line at an angle of 39°, being tangents to a circle of 18 feet diameter, and on an average head and dip of 14 feet—11 feet head and 3 feet dip—enter the water at an angle of 34°, and leave it at an angle of 72°. The wheel makes $4\frac{1}{4}$ revolutions a minute. When working to its full extent, the wheel is capable of discharging 213 tons per minute. This wheel has been provided with a movable breast as at the other station. The engine for driving the wheel is similar in character to that at the Hundred Foot Station, and was altered and adapted for working with a higher pressure of steam in a manner similar to the other. The cost of alterations at the two stations was over 6000*l*. The estimated capacity of the two wheels at the maximum dip is 410 tons per minute. This is equal to a discharge of water due to a continuous daily fall of 0·17 inch of rain. In the year 1883, which was a very wet season, the engines ran as follows:—

		Hundred Foot Engine.	Ten Mile Engine.
Total hours run		2288	2280
Coal consumed	tons	691	589
Average dip of the scoops	feet	3·30	3·08
Greatest ,, ,,	,,	5·33	4·58
Average head	,,	13·80	11·16
Greatest ,,	,,	17·2	13·4

The estimated discharge, calculated with the average dip of the scoops given above, is 122·12 tons per minute lifted 13·80 feet, equal to 114·40 horse-power of water lifted, with a coal consumption of 5·99 lb. per horse-power of water lifted for the Hundred Foot engine, and 128·55 tons lifted 11·16 feet, equal to 97·38 horse-power, with a coal consumption of 5·93 lb. per horse-power for the Ten Mile Station. Taking the two wet years, 1881 and 1883—1882 being omitted, as during this time the machinery was under alteration—the cost of lifting the water was as follows :—Coal, 717*l.*; attendance and other expenses, 203*l.* The area drained being taken at 35,000 acres, this gives 12·62*d.* per acre per annum for working expenses. The average height to which the water was lifted at the two stations being taken at 11½ feet, gives 1·10*d.* per acre per foot of lift, or, for coal only, of 0·85*d.* per acre, coal costing about 17*s.* per ton.

WHITTLESEA MERE.—This pumping station is in the Middle Level, in the county of Huntingdon, and contains about 6000 acres. The Mere originally was a large lake or morass, which produced nothing but reeds and wild fowl. This, with the surrounding fen, was embanked and drained by steam power by the proprietor, Mr. Wells, in 1851-2, being the first instance in this country where the centrifugal pump was applied to this purpose; the results obtained with the Appold pump at the trials of this machine at the Exhibition of 1851 demonstrating its suitableness for the purpose. The engine then erected was of 24 nominal horse-power, driving a double inlet horizontal spindle Appold centrifugal pump, 4 feet 6 inches diameter, with an average velocity of 90 revolutions a minute, equal to 1431 feet per minute; the lift at that time being from 4 feet to 5 feet. The pump was driven by a double-cylinder steam engine, with steam at 40 lb. pressure, and vacuum 13½ lb. It raised 15,000 gallons—67 tons—per minute to a height varying from 2 feet to 5 feet. The total cost was 16,000*l.*, of which about 2000*l.* was for the machinery. The general arrangement of the pumps is shown by the sketch (Plate 6). The pump discharged into Bevil's river, a branch of the Nene, which forms a part of the great Middle Level system, the outfall of the main drain being into the River Ouse, at St. Germains, 30 miles distant. The soil of this district is almost entirely peat, to a depth of from 15 feet to 18 feet. After the drainage operations had been at work some time, the surface of the land gradually lowered, owing to the waste and shrinkage of the peat. In July 1857 the level of the water in the drain was 5 feet above datum; and in August 1860 it was reduced to 3 inches above datum. At the present time the surface is about 8 feet lower down than it was thirty-two years ago, when the district was first drained. The pump was twice lowered during the twenty-six years it was at work, until the lift was increased to over 9 feet, thus demonstrating the peculiar facility this class of machine has to meet such an occurrence. Owing, however, to the increased lift, and the altered circumstances of the district, it became necessary to increase the pumping power. The average lift now is about 7 feet, rising frequently to 9 feet 6 inches, and even higher in heavy floods. In 1877 the old engine and pump were removed, and the fan of the pump may now be seen at the Museum at South Kensington in almost perfect condition. Messrs. Easton & Anderson erected in their place a high-pressure compound condensing beam engine, with expansion gear, of 65 nominal horse-power, making about 36 revolutions a minute with 60 lb. steam. The boilers consist of one single-flued and one double-flued Cornish boiler. The pump, which is placed in a well outside the engine-house, is driven by a double set of motions, the first set consisting of a toothing on the fly-wheel driving a pinion, which actuates a horizontal shaft for driving a wheel geared into a bevil-wheel on the vertical shaft of the pump. This is hung by an onion-bearing to a cast-iron frame bolted to the top of the pump-well, which is formed with a wrought-iron cylinder fixed in the centre of the sluice connecting the main drain with the river. This cylinder was used as a convenient mode under existing conditions of forming the pump-well, and reduced the first cost by avoiding the necessity for building a brick well. This sluice is 12 feet wide on the inlet side and 6 feet on the delivery side. The fan is a single inlet fan of 6 feet diameter by 16 inches deep, and is speeded to run up to 104 revolutions a minute when on a lift of 11 feet. The quantity of water delivered is 96 tons per minute, or on a lift of 7 feet 6 inches, with a speed of 96 revolutions of pump, 155 tons per minute. The engine and boiler are contained in a brick building. The chimney shaft is 53 feet high, and 3 feet diameter at the top inside. The cost of the machinery was approximately 3500*l.*, plus the value of the old machinery.

BURNT FEN.—This district is situated in the South Level of the Bedford Level, in the county of Norfolk, and is entirely fen land. The area drained by the pumps is 15,000 acres. There are two pumping stations, about 4 miles apart—one at the Fish and Duck, on the south side of the district, discharging into the River Lark about 3 miles above its junction with the Ouse; the other on the north, discharging into the Little Ouse about 2 miles above Brandon Creek Bridge. The main drains between the two stations are in connection, so that the water can run to either station. These pumping stations are about 8 and 15 miles respectively above Denver Sluice, where are self-acting doors, which shut against the tide at the time of high water. The lift at the north station is rather the highest, the average of the two stations being about 10 feet 6 inches, rising in heavy floods to 16 feet. The north station consists of a scoop wheel 34 feet 6 inches in diameter, with scoops 4 feet 9 inches long by 2 feet wide, motion being given by one engine of 40 nominal horse-power. The wheel is driven by a condensing engine of the old marine side-lever type, having the beam below the cylinder. The piston has 3 feet 6 inches stroke, and makes 28 revolutions of the engine to $5\frac{1}{2}$ of the wheel. The working pressure of the steam is 15 lb. on the inch. The station at the Fish and Duck was provided, until recently, with a scoop-wheel; but as, owing to the subsidence of the peat, the surface has settled in this district 4 feet 6 inches since the beginning of the present century, it was necessary to provide more efficient machinery; and under the advice of Mr. Carmichael, the superintendent of the works in the South Level, the scoop wheel and engine were replaced by a centrifugal pump. The new engine is of the horizontal tandem type, high-pressure compound condensing, fitted with expansion gear, 60 nominal horse-power, the cylinders being 18 inches and 30 inches in diameter, with 3 feet stroke, provided with variable expansion valve working on the back of the high-pressure valve. Steam is provided by three Lancashire boilers, 25 feet long by 7 feet diameter; the working pressure being 65 lb. Only two of the boilers are in use at the same time. The engine makes 70 revolutions with steam at 65 lb. in the boiler, and cut off in the small cylinder at half of the stroke, the pump making at the same time 105 revolutions with a lift of 14 feet per minute, and delivering 120 tons. The case of the pump is 9 feet 6 inches diameter, situated in a well immediately outside the wall of the engine-house. This well is 9 feet 10 inches in diameter; the diameter diminishing below the pump to 6 feet. The outlet for the discharge is 9 feet 6 inches above the centre of the pump, and is 5 feet 6 inches high by 3 feet 6 inches wide. The pump is driven by a bevel wheel geared into a bevel pinion on the crank shaft, which is 11 feet long. The fan is single, made of gun-metal, 6 feet diameter by $12\frac{1}{2}$ inches deep at the periphery, with a short suction-pipe attached to the case below the disc. The spindle is suspended by an onion-bearing, supported by a girder across the top of the cylinder of the pump-well. When the pump is working it is found that little weight is carried by the onion-bearing, as the disc is so arranged that the water entering it supports the moving parts. The pump is calculated to lift the following quantities: —121 tons at 9 feet; 115 tons at 10 feet; 109 tons at 11 feet; 104 tons at 12 feet; 100 tons at 13 feet; 96 tons at 14 feet; 92 tons at 15 feet; 89 tons at 16 feet. These quantities were exceeded at the trials of the pump. The engine-bed occupies a space of 30 feet by 5 feet 6 inches. The engine and pump were supplied by Messrs. Hathorn, Davey & Co., of Leeds. The contract price, including the well and fixing in the old building, the makers taking the old engines, was 2700l. A drawing showing the arrangement of the pump and engine will be found in 'The Engineer,' vol. lvii., February 1884, and an enlarged view of the pump is now given on Plate 6. Careful observations have recently been taken by Mr. Carmichael as to the consumption of coal by this engine under ordinary working conditions, the quantity of water delivered being ascertained by measuring the quantity passing through the outlet drain. With a lift of 11 feet the quantity of water discharged was 120 tons per minute, with a consumption of 3 tons of Derbyshire coal in twelve hours. This is at the rate of $6\frac{1}{4}$ lb. per horse-power of water lifted per minute. The quantity of oil used for lubricating is at the rate of 1 gallon in twelve hours. The consumption of coal in this district has varied during the last twenty years from about 250 tons to 1000 tons in a year according to the rainfall; the average cost for the years 1881–3 (coal being then about 15s. per ton) was for coal, 674l.; attendance, oil, &c., 252l.; total, 926l. Taking the average lift for both stations at $10\frac{1}{2}$ feet, this is equal to 14·81d. per acre, or per acre per foot of lift, 1·42d.; or for coal only, 1·02d. During this time both scoop wheels were in operation. The main drain, which brings the water to the pump, is 20 feet wide at the bottom, with slopes of $1\frac{1}{2}$ to 1. The average depth of water when pumping is going on varies at starting from 5 feet 6 inches to 3 feet at leaving off; the surface inclination also varying from $2\frac{1}{2}$ inches per mile to 4 inches.

PRICKWILLOW.—This pumping station is for the drainage of a large district in the South Level, being part of the Great Bedford Level, in the county of Cambridge. The taxable area of the district is about 11,000 acres; but the area of land actually drained by the engines at Prickwillow is about 25,000 acres, the drainage of a large area of higher land bordering on the Fens finding its way into this Fen drainage system. The water is lifted by both engines into the River Lark, about fourteen miles above Denver sluice, where the river discharges into the tidal stream from the same main Fen drain, which is 20 feet wide, with slopes $1\frac{1}{2}$ to 1. The depth of water at starting the engines is generally about 6 feet 6 inches, decreasing to 4 feet 6 inches after the pumping has been going on. Since the erection of the new engine and pump this drain has been found to be too small to keep up a full supply, the inclination on the surface being at the rate of 6 inches in a mile, which is greater than should be the case in a large main engine drain. The height the water has to be raised on an average is 10 feet, rising as high as 17 feet in high floods in the river. Steam power was first applied to the drainage of this district in 1832, a 60 horse-power low-pressure condensing engine being then erected by the Butterley Company to drive a scoop wheel 33 feet 6 inches in diameter; and this engine, with the aid of numerous wind engines previously in use, and retained as auxiliaries, preserved the district from injury fairly well. The continuous subsidence of the surface of the land, and the increased height the water rose in the river, due to the rapidity with which floods now come down from the uplands, rendered this drainage power inadequate. It was found by experience that, owing to the constant variations in the levels of the water, both in the main drain and in the river, the scoop wheel became so water-logged and unwieldy, and the loss by leakage so increased by the great head of 10 feet to 13 feet, against which it frequently had to work, that, notwithstanding the great prejudice which all Fen men have in favour of the scoop wheel, Mr. Carmichael, the superintendent of the South Level, advised the Commissioners to adopt another form of machine which would adapt itself automatically to the variations of lift, and which, under the varying circumstances of the discharge, would absorb the whole power of the engine to the best advantage, and for this purpose he selected one of the Appold type, which, although they had been in use for some time in other parts of the Fens, were as yet untried in the South Level. The new engine and pump were intended to relieve the old engine of the greater part of its duty, more especially in times of excessive floods, and to drain out the water to a lower level than was practicable with the scoop wheel. The new machinery was erected by Messrs. Easton & Anderson, under Mr. Carmichael's direction. The engine is a 60 nominal horse-power compound condensing beam engine, supplied with steam at 65 lb. pressure by two Lancashire boilers. The high-pressure cylinder is 15 inches, and the low-pressure 25 inches diameter, with 4 feet

Pumping Stations. 123

6 inches stroke. The pump is of the vertical spindle pattern, with single inlet, with balance fan 5 feet 4 inches diameter and 1 foot 2 inches deep, placed at such a level that the lowest water in the drain will cover it. The inlet is 2 feet 8 inches diameter, formed on the lower side only, special provision being made for balancing, the weight of the column of water above the fan being balanced by the fixed inlet piece, which also serves to steady the lower end of the fan spindle. The meeting faces between the fan and the fixed case are both turned in the same direction, so that wear as it takes place can be taken up simply by lowering the fan spindle by means of an adjustment provided for the purpose. To take up the momentum of the water issuing at great speed from the fan, patent guide curves were fitted, which turned the water gradually into the vertical direction, and at the same time assisted to bring it to rest. In this particular instance these guide curves were not found to be of much avail, as when the river was very low, the delivery was lower than the top of the blades, and consequently there was a churning action going on with the water in the well, which caused vibration in the spindle. They were, therefore, removed. The pump is placed at the bottom of a brick well, in one side of which is the outlet passage 4 feet wide by 4 feet 6 inches high, fitted with self-acting doors, and communicating with a cast-iron outlet pipe 4 feet 6 inches diameter and about 68 feet long. The upper end of the fan spindle hangs in an onion bearing, and is driven by a pair of bevel wheels from a horizontal shaft which passes into the engine-house, on which is a pinion driven by annular gearing, bolted to the rim of the fly-wheel of the engine. The pump is calculated to lift 95 tons per minute at 8 feet lift, 88 tons at 9 feet, 83 tons at 10 feet, 78 tons at 11 feet, 74 tons at 12 feet, 71 tons at 13 feet, 68 tons at 14 feet, 65 tons at 15 feet. The cost of the machinery, including engine, pump, and two boilers, was 3853*l*. The buildings, engine-house, boiler-house, pump-well, chimney-base, piling, and concrete, cost about 1064*l*. At the trials which took place when the new engine was started it was found that the old engine indicated 103·33 horse-power when delivering the water to a height of 9·78 feet; the new engine when indicating 106 horse-power delivered 75·93 tons to a height of 10·84 feet; the coal consumption was at the rate of $2\frac{3}{4}$ cwt. per hour, and as compared with that of the old engine in the proportion of 3 to 5. At a subsequent trial a weir 13 feet wide was placed across the outlet drain, the difference of level between the water in the inlet drain

and the weir at starting was 8 feet 9 inches; with the scoop-wheel the depth of water over the weir was 12 inches, with a lift of 9 feet 6 inches; with the pump, the lift being 10 feet, the depth of water over the weir was $13\frac{1}{2}$ inches. The lift being increased about 3 feet, the depth of water over the weir was 4 inches less with the scoop wheel than with the pump. At the trials that were made, the new engine, indicating 106 horse-power, 75·93 tons of water were lifted by the pump 10·84 feet, equal to 56 horse-power of water lifted, or an efficiency of 52·79 per cent. The old engine, indicating 103·33 horse-power, the wheel lifted 71·45 tons to a height of 9·78 feet, equal to 47·43 horse-power of water lifted, or an efficiency of 46 per cent.; the coal consumed by the new engine was at the rate of $2\frac{3}{4}$ cwt. an hour, or 5·50 lb. per horse-power of water lifted per hour. In ordinary working at the present time the consumption is at the rate of 5 tons in 30 hours for a lift of from 11 feet to 12 feet. Taking the horse-power as before at 56, this gives 6·66 lb. per hour; or, if the work be taken at 74 tons lifted 11 feet 6 inches high, a horse-power of 58·45, and coal consumption of 6·39 lb. The old engine and wheel consumes 6 tons of coal in 24 hours; if the horse-power be taken at 48·12 as before, this gives 11·64 lb. an hour. The cost of this pumping station, including both machines, on an average of the three years, 1881–83, for coal, oil, attendance, &c., was 625*l.*, of which 483*l.* was paid for coal, which represents about 644 tons. This is equal to a cost per acre for land drained of about 6*d.*, or, taking coal only, 4·62*d.*, and taking the average height the water has to be lifted at 9 feet 6 inches, this is equal to 0·8*d.* for all expenses, and 0·62*d.* for coals only per acre per foot of lift.

THE UPWELL, OUTWELL, DENVER, AND WELNEY south district is situated in the Middle Level in Norfolk, being part of the Great Bedford Level. This district was originally drained by scoop wheels driven by windmills. The quantity of land which is drained by the two wheels is about 9000 acres. The pumping station is at Nordelph, about three miles from Downham. It was anticipated that the construction of the new Middle Level drain in 1846 would do away with the necessity of pumping the water off the district, but experience showed that this was not the case. The height to which the water had to be raised was however reduced from about 9 feet to 4 feet. In order, therefore, to thoroughly drain this district, the commissioners determined to provide better appliances for raising the water than those hitherto in use. Tenders for pumping machinery were advertised for, and that of Messrs. Appleby & Co. was accepted. The new machinery, the arrangement of which is shown on Plate 6, was erected in 1877, and consisted of a scoop wheel 24 feet in diameter by 4 feet wide, and, according to the maker's calculation, capable of delivering 3500 cubic feet (98½ tons) per minute to a height of 4 feet, equal to 26·51 horse-power of water lifted. The wheel makes five revolutions a minute, equal to a speed of 6·27 feet per second at the periphery. It is constructed principally of wrought iron. The scoops, eighteen in number, 10 feet long, and $\frac{3}{16}$ inch thick, are curved and shrouded by wrought-iron plates, and are connected to the wheel by curved arms, 2 inches by 2 inches by $\frac{3}{8}$ inch. The sides are $\frac{3}{16}$ inch thick at the periphery to $\frac{3}{8}$ inch at the centre. An adjustable curved shuttle is provided at the inlet to the wheel by which the admission of the water is regulated. This shuttle is supported at the top by two arms which project and clasp the axle of the wheel. Part of the pressure of the water against the shuttle is thus brought to bear on the [axle, causing considerable friction. The sill over which the water is delivered is curved to the radius of the wheel. The wheel is keyed on to a wrought-iron shaft 9 inches in diameter, which runs in adjustable gun-metal bearings. On one side of the wheel is bolted a geared wheel made in segments 20 feet in diameter, of 3-inch pitch and 6-inch face, and into this works the pinion on the engine crank shaft. The engines are of 40 nominal horse-power, of the horizontal high-pressure, compound condensing type. The high-pressure cylinder is 10 inches in diameter and 20 inches stroke. The low-pressure cylinder is of 20-inch diameter and 20-inch stroke. The low-pressure cylinder and condenser are on one base, the air-pump being fixed in the chamber of the condenser. The high-pressure cylinder is placed on a separate base parallel with the other cylinder. The fly-wheel is 9 feet in diameter. Steam for the engines is generated in two Cornish boilers, 20 feet long by 5 feet diameter, fitted with Galloway tubes, the safety valves being weighted to a pressure of 80 lb. of steam. The engines and boilers are contained in a brick building. The chimney is of brick, built square, 60 feet high. The contract price for building and machinery was 2680*l*., of which 700*l*. was for the buildings, chimney, and casing for wheel. This is equal to 74·68*l*. per horse-power of water lifted for the machinery, and 26·40*l*. for the buildings, together 101·08*l*. This is the only wheel in the Fen-land that has curved scoops. The head and dip of this wheel in ordinary floods are about 8 feet 6 inches, the relative proportions of each varying as the water lowers in the inlet or rises in the outlet drain. As an average the dip may be taken at 4 feet 6 inches and the head at 4 feet. With the wheel making five revolutions a minute, and allowing 20 per cent. for slip of water and leakage,—and this deduction is borne out by the quantity of water flowing down the engine drain,—the discharge is equal to 4305 cubic feet—120 tons—a minute. The quantity of coal consumed for this discharge, with 4 feet head, is about two tons in twelve hours, equal to 11·440 lb. per horse-power per hour of water lifted. By the side of the engine-house stands one of the old windmills which is still used to drive a scoop wheel 20 feet in diameter and 2 feet wide, and which when there is sufficient wind assists in raising the water from the district. When both steam and wind engines are at work, the quantity as given above is about equal to the discharge of a continuous fall of $\frac{1}{4}$ inch of rain in twenty-four hours over the area of 9000 acres, of which the district is comprised.

GLASSMOOR.—This is a district also in the Middle Level, consisting of about 6000 acres of Fenland. It discharges its water into one of the main drains of the Middle Level system about twenty-seven miles from the outfall sluice in the Ouse. The average lift is about 5 feet, rising occasionally in floods to as much as 8 feet. The machinery consists of a pair of 15 nominal horse-power high-pressure condensing vertical engines; the crank shaft is carried on cast-iron A frames, the fly-wheel working inside these and toothed into a horizontal bevel wheel attached to the vertical shaft of the pump, which is placed in a well immediately under the engines. The steam cylinder, condenser, and pump, are outside the frame, the latter being worked by a rocking beam, one end of which is connected with the piston-rod, and the other to the floor. Steam is supplied by two Cornish boilers, the working steam pressure being 40 lb., the engines making forty-seven revolutions a minute, and the pump at this speed 116. The culvert for connecting the pump-drain with the river passes under the engine-house, the pump well being in the centre. The pump has a 4-feet fan, 1 foot $1\frac{1}{4}$ inch deep. The engine and boilers are contained in a well-designed building of white bricks with red arches and dressings, which present a very pleasing appearance; the chimney is about 70 feet high. This machinery was erected about twenty-four years ago by Messrs. Easton and Amos, and has stood and worked during that time without any material repairs. The framework of the engines only occupies a space about 6 feet square, and with the pump being placed under this frame, the cost of foundations was reduced within a small compass. The long time that the engines have run is considered to be partly due to their being of the vertical type, the wear and strain on the cylinder being less than in a horizontal engine, and a settlement in the foundations having less effect on the working of engines arranged as these are. This station is an illustration of the suitability of the centrifugal pump for Fen drainage, and shows that pumps, equally with scoop-wheels, will run for a great number of years without accident or repair. The consumption of coal for the three years 1881–83, averaged about 60 tons a-year, equal to 100 acres per ton of coal. Taking the cost of coal at 15*s.* a ton, this is equal to 1·80*d.* per acre for coal, or taking the lift at 5 feet, ·36*d.* per foot of lift.

MESSINGHAM DISTRICT, LINCOLNSHIRE.—The description of this pumping station is given as an illustration of works carried out in an inexpensive manner for the drainage of a small district where it was not considered desirable at the time to incur an outlay sufficient to put up works of a more permanent character. Owing to the difficulty in the way of obtaining a site for the buildings, the engine-house and pump are erected on piles over the main drain near the outfall sluice, and the discharge pipe is carried under the highway to the river. The engine and pump are situated at Butterwick, in the north-west part of Lincolnshire, and were erected by Mr. C. L. Hett, of Brigg, under the direction of Mr. Alfred Atkinson for the Commissioners of Sewers. The extreme range of a spring-tide in the Trent at this place is about 18 feet, and the consequence is that the scoop wheels, which have hitherto been almost exclusively used for drainage purposes in that part of the country, can only work for four or five hours each tide, or in some cases for even less. It was therefore determined to use a centrifugal pump. An illustration of the general arrangement of the machinery will be found in Plate 2.

The district drained includes 3250 acres adjoining the river Trent, and comprises some very low-lying land. Previous to the erection of the pumping machinery, the drainage had been by gravitation through outfall sluices, the sills of which are about level with ordinary low-water in the river. This system was found to be inefficient in wet seasons when good drainage was of the greatest importance. When there was much rain falling in the upper districts drained by the Trent, the sluice doors were kept closed, sometimes for days together, during which the rainfall in the district accumulated in the drains and ultimately overflowed the low grounds. The result during wet seasons was most disastrous to the agricultural population of the district. As the only means of relief, the Court of Sewers determined to erect pumping machinery. The cost of the works was defrayed by a tax on the land; and, as many of the contributors had become greatly impoverished by several successive bad harvests, the greatest economy had to be exercised. The pumping machinery was first started in March 1882, and has since been working satisfactorily. Only one engine and pump were erected, but it was intended to fix a duplicate set at a future time.

The centrifugal pump is of a pattern described as "Hett's Improved Accessible," with suction and delivery pipes 21 inches in diameter. It is so arranged that the side of the case can be removed, and the interior inspected, or the disc removed without breaking any pipe joints or connections. The pump is charged by means of a steam-jet exhauster. The delivery-pipe has a submerged bell mouth, and is fitted with a sluice valve near the pump. The pump is driven by a belt from a double cylinder semi-portable engine, fitted with Hartnell's automatic expansion valve gear. The quantity of water this pump was estimated to discharge was 10,000 gallons (44·64 tons). The lift varies from a few feet to about 12 feet at ordinary spring-tides, increasing to 14 feet and even 16 feet at high tides. The total cost of the machinery, with an engine-house large enough to contain two pumps, was 1432l.

During the excessively wet season of 1882–3 a great strain was thrown on this machinery; owing to the fact of the second pump not being provided, it was not adequate to the work required. Notwithstanding this disadvantage, the machinery has proved of the greatest benefit to the district concerned. On 23rd October, 1882, occurred one of the heaviest rainfalls known in the neighbourhood; and almost simultaneously the highest recorded tide in the River Trent. The district is bounded on the north and south sides by streams draining large tracts of upland, which were so surcharged as to overflow the floodbanks; and at the same time a small breach occurred in the Trent bank. All the water that gained access to the district by these means had to be pumped out, in addition to the rainfall, because the fresh in the Trent was too great to allow the outfall sluices to act. On this occasion the machinery was running for 197 consecutive hours, stopping only for oiling now and then. It was kept at work throughout the extraordinary tide mentioned above. The scoop wheels in the neighbouring lands were at the same time so completely drowned out that they could not be used for many days after—the result being serious and long-continued inundations.

REDBOURNE, LINCOLNSHIRE.—This is another example of an installation for the drainage of a small area of land, where it was considered desirable to avoid the cost of foundations, and that the first outlay should be as small as practicable. The area drained consists of 800 acres of car land adjoining the river Aucholme in North Lincolnshire. The surface of the land is below the level of floods in the river, and there is very great leakage through the banks. It was estimated that the rainfall and leakage together would require that the quantity of water to be pumped would be equal to an amount due to a continuous rainfall of half an inch in twenty-four hours, or 21·10 tons per minute, and that this would have to be lifted an average of 7 feet high, equal to 10 horse-power in water lifted.

The installation consisted of one of Hett's side-opening centrifugal pumps, having an 18-inch inlet, driven by a belt from a 14 nominal horse-power semi-portable double-cylinder engine, placed in a brick and tiled shed built on the adjacent bank. A cast-iron outlet, 18 inches in diameter, was carried through the bank to the river, the outlet being placed below the lowest point to which the water was likely to fall. The pump is primed by a steam-jet exhauster, and the feed-tank was also fitted with a jet-pump. The total cost was as follows :— Engine and pump, 641*l*.; brick and tiled shed and other work, 211*l*.; together, 852*l*.; equal to 85·20*l*. per horse-power in water lifted.

The machinery was supplied and erected by Mr. Charles L. Hett, of Brigg, for the Duke of St. Albans, the owner of the land, under the direction of the author.

LEVEL OF HATFIELD CHACE, BETWEEN DONCASTER AND GOOLE.— This district, which lies on the borders of Lincolnshire and Yorkshire, and comprises a tract of low-lying land of about 18,000 acres, very similar in quality to the Fens of Lincolnshire and Cambridgeshire, has undergone very considerable improvement, both in natural and artificial drainage, during the present century. The original system of natural drainage was established by Vermuyden above two hundred years ago, the main features of which were the cutting of large outfall drains, with suitable sluices, having their outlets into the river Trent at Keadby, Althorpe, and Owston Ferry in Lincolnshire, also the cutting of what is known as the Dutch River, which empties itself into the river Ouse at Goole. The Level is separated into two districts by the ancient River Torne, which brings upland waters for many miles, and discharges the same at Althorpe.

South District, Wroot, or Bull Hassocks Engine.—The area of this district is 7270 acres, the average level of the surface of the land being only a few feet above low-water mark in the Trent, consequently the natural drainage was totally insufficient in wet seasons. Upwards of forty years ago a steam-engine was erected at a place called Little Hirst, about $3\frac{1}{2}$ miles from the outfall at the Trent; but experience showed that it was placed too far from its work, and in 1857 it was removed to its present position, Bull Hassocks, near Wroot. The engine was not new when purchased, having been constructed for marine purposes. It is a side-lever engine of 40 horse-power nominal, equivalent to 70 water horse-power. The scoop wheel is 30 feet diameter and 2 feet $11\frac{1}{2}$ inches wide, and works at the rate of $4\frac{1}{2}$ revolutions per minute, with an average lift of 5 feet.

North District, Dirtness Engine.—This district contains 10,660 acres. The entire works of drainage were previous to the year 1862 vested in the "Trustees of Decreed Lands"; but in that year, by Act of Parliament, they were incorporated under the title of the "Corporation of the Level of Hatfield Chace," and twelve commissioners were appointed in the usual manner. Powers were taken to improve the drainage of the entire Level, and to erect machinery for the north district at Dirtness, about two miles from the town of Crowle. The new engines were built in 1864–65 by Messrs. Watt and Co., of the Soho Works, Birmingham, and comprise two compound condensing beam engines, each 50 horse-power nominal. The high-pressure cylinder is 20 inches diameter, with a stroke of 4 feet $4\frac{1}{2}$ inches. The low-pressure is 35 inches diameter, with 6 feet stroke. The two engines are coupled at an angle of 90° to a crank shaft carrying the fly-wheel, and a pinion which gears into a wheel with wooden cogs and shaft passing through the engine-house wall, and carrying a pinion gearing into teeth cast with the rim of the scoop wheel. Steam is supplied by four double-flued boilers, 20 feet long by 7 feet diameter, working to a pressure of from 20 lb. to 30 lb. steam. The scoop wheel is 33 feet 3 inches diameter and 6 feet wide, and is capable of raising and delivering 12,000 cubic feet of water 7 feet

Pumping Stations. 131

high per minute, equal to 159 water horse-power. It contains 36 scoops, with a radial length of 7 feet 10 inches each. These enter the water at an angle of 13° and leave it at 31°. Extreme dip, 7 feet; average dip, 4 feet; and average lift, 4 feet 9 inches. Number of revolutions of engines per minute, 26; and those of the scoop wheel, 4. The wheel has 8 spokes each in one casting of the width of the wheel, with three rims bolted to the spokes, and each carrying a set of oak start-posts 7 inches by $3\frac{1}{2}$ inches at the rim, and $4\frac{1}{4}$ inches by $3\frac{1}{2}$ inches at the circumference. Each set of start-posts is held together by two wrought-iron rings $2\frac{1}{2}$ inches by $\frac{1}{2}$ inch, one in the middle and the other about 4 inches from the water end of the start. The floats are of 1-inch fir, and planking is also carried round the wheel at the inner end of the floats. The wheel contains in planking 166 cubic feet of fir timber, equal to about $2\frac{1}{2}$ tons, and in oak start-posts 115 cubic feet, weighing about $2\frac{1}{2}$ tons. The buildings cover 3990 superficial feet of ground, the boiler-house being 41 feet square; the engine-house is 52 feet 6 inches by 28 feet; and the wheel-house, 52 feet 6 inches by 16 feet.

The cost of the engines, boilers, and scoop wheel was 4340*l*., and of the building, 4547*l*. Taking the horse-power at 159 W.H.P. this gives 26·30*l*. for the machinery, and 28·60*l*. for the buildings; together, 55·90*l*. per horse-power in water lifted.

WEXFORD HARBOUR RECLAMATION WORKS, IRELAND.—A large area of land was reclaimed from Wexford Harbour by embankments. From the level of this land, as compared with the water in the harbour, it was necessary to use steam power for the drainage. The reclamation is divided into two districts, termed respectively the North, containing 2489 acres, and South, containing 2410 acres, each having a separate pumping station; that for the south side being a scoop wheel and for the north a centrifugal pump. The water pumped off is exclusively rainfall, which in ordinary seasons amounts to $45\frac{1}{2}$ inches. It was calculated that three-fourths of this would have to be pumped, the remainder being absorbed by the vegetation or taken up by evaporation. Spring tides in the harbour rise 5 feet, and neap tides 3 feet. The pumping station for the south reclamation is situated at Drinagh, and consists of a scoop wheel driven by a condensing beam engine. The engine has one 36-inch cylinder, with 6 feet stroke. A variable rate of expansion can be given by sliding the cams which work the steam valves. The engine runs at the rate of 25 revolutions a minute. Steam is supplied by

The Drainage of Fens and Low Lands.

two Cornish boilers, 21 feet 6 inches long, the working pressure being 13 lb. on the inch. The scoop wheel is 40 feet diameter by 10 feet wide, the scoops being 3 feet long. The wheel makes $4\frac{1}{2}$ revolutions a minute when the engine is making 25, the velocity at the periphery being 9·42 feet per second. At this pace the wheel was calculated to raise 170 tons of water per minute. The water approached the wheel with a velocity of 9·42 feet per second—equal to that due to a fall of 1·4 feet, leaving the net calculated lift 8 feet. The useful effect at the trials was found to be 68·2 per cent., leaving the loss from all causes, including the engine, 31·8. The fuel consumed was found to be $4\frac{1}{2}$ lb. per indicated horse-power per hour. The total cost of the machinery for this pumping station was 5000*l.*—equal to 2·08*l.* per acre drained. Taking the greatest duty at 170 tons lifted 11 feet per minute, and the total cost at 5000*l.*, this gives about 40*l.* per horse-power of water lifted for the machinery, the contract price for the wheel being 760*l.* The wheel weighs $34\frac{1}{4}$ tons, and is carried by a cast-iron hollow shaft 13 inches diameter, working in brass bearings 12 inches diameter by 16 inches long. It has three cast-iron spoke centres 6 feet diameter; twenty flat wrought-iron spokes radiated 14 feet from the periphery of the centres. A ring of flat iron, 34 feet diameter, connects the ends of the arms. To this ring, and a smaller one, 28 feet diameter, are riveted flat-iron float spokes, bent to give the scoops the proper rake. The spur wheel is 31 feet $6\frac{1}{2}$ inches diameter, $10\frac{1}{2}$ inches wide, and $3\frac{1}{2}$ inches pitch. The spur pinion, 5 feet $3\frac{1}{2}$ inches in diameter, is keyed on to the fly-wheel shaft, and gears into the annular wheel 7 feet below its centre on the discharging side, by which arrangement it was considered that the weight of the water would be borne directly by the pinion. There are forty scoops, each 9 feet $11\frac{3}{4}$ inches wide by 3 feet deep, drawn tangents to a circle 13 feet diameter, and formed of 3-inch Memel planks, grooved and tongued, and secured by hook bolts to the float spokes. The clearance between the wheel and the masonry is from $\frac{1}{4}$ inch to $\frac{3}{8}$ inch. The water is delivered over a crest 4 inches broad, 11 feet higher than the bottom of the race under the wheel, and 3 feet 6 inches below high-water spring tides. Water is admitted to the wheel by cast-iron sluices. The channel from the wheel to 8 feet outside the inlet is level, 10 feet 7 inches wide on wheel side of sluice, and 11 feet 7 inches on outside of sluice, the sill of which is 3 inches higher than the bottom of the channel.

The pumping station for the North Reclamation was erected after the scoop wheel had been in operation some time, and it was determined to adopt a centrifugal pump as the more effective machine under all circumstances, its great advantage over the wheel being its adaptability to varying lifts and less cost of foundations. The pump used was one of Appold's type, supplied by Messrs. Easton and Anderson. It is self-contained in a cast-iron frame, with galvanised iron fan, 4 feet diameter by 15 inches deep, with diaphragm in middle of its depth, and revolving in cast-iron case. Two suction pipes conduct the water to above and below the fan, which is carried by a vertical spindle making 133 revolutions a minute. Motion is given to the spindle by a bevel pinion with 43 teeth geared into mortice bevel fly-wheel, with 114 teeth keyed on to the crank shaft. The pump is driven by a pair of direct-acting condensing engines, having cylinders $18\frac{1}{4}$ inches diameter, with 2 feet stroke; steam being used at pressure of 50 lb., with high degree of expansion, and supplied by a Cornish boiler 22 feet long by 6 feet 6 inches diameter. The consumption of fuel at the trials was found to be $4\frac{1}{2}$ lb. per I.H.P. Allowing an efficiency of ·54, this is equal to 8·33 lb. per W.H.P. The cost of the pump and engines complete was 1850*l*. The buildings, culverts, and foundations cost 2725*l*.; together, 4575*l*., equal to 1·84*l*. per acre drained, and 37*l*. per horse-power of water lifted for the machinery, and 54*l*. 10*s*. for the buildings; together, 91*l*. 10*s*. per W.H.P. Trials of this pumping machinery were made soon after the erection, steam in the boiler being from 30 lb. to 35 lb., and the mean pressure on the piston varying from 14·79 lb. to 17·02 lb., the engine making $47\frac{1}{2}$ to $54\frac{1}{2}$ revolutions, and the lift—that is, the difference of level of water in mouth of inlet culvert and in the outlet at engine-house—varying from 6 feet 2 inches to 10 feet 2 inches. The indicated horse-power varied from 43·6 to 58·8, and the horse-power of water lifted 24·07 to 31·67, giving a mean effective result of 53·8 to 55·2 per cent., or allowing one-sixth of indicated horse-power as the resistance of the engines, the mean duty of the pump was 67 per cent. The particulars of these two stations are taken from a paper by Mr. W. Anderson in the 'Proceedings' of the Institution of Civil Engineers in Ireland for 1862 (vol. vii.), in which will be found the full details of the trials and drawings of the scoop wheel, &c. For the three years 1881 to 1883, the average rainfall at these stations was 40·34 inches. The scoop wheel, which drains 2410 acres, worked on an average 1800 hours each year, and the engines consumed about 350 tons of coal. The pump on the North Reclamation, which drains 2489 acres, ran on the average 1516 hours, and the engines consumed 215 tons of coal each year. Taking the average lift throughout the year in both cases as 5 feet 6 inches, and coal at 18*s*. per ton, this gives 26·3*d*. per acre per annum for the scoop wheel, and 18·45*d*. for the pump for coal only, or per acre per foot of lift 4·82*d*. and 3·35*d*. respectively.

FERRARA MARSHES, NORTH ITALY.—This pumping station contains one of the largest combinations of centrifugal pumps for the drainage of land yet supplied. The machinery was erected in 1873 by Messrs. J. and H. Gwynne, for pumping the water from the Ferrara Marshes in North Italy. The reclaimed land extends over an area of nearly 200 square miles, and the work done by the pumps consists in raising a little over 2000 tons of water per minute for a mean lift of 7 feet 6 inches—the maximum being 12 feet—and delivering it into the river Volano, at Codigoro. The machinery consists of four pairs of centrifugal pumps having vertical discs, each set driven by a pair of compound engines. Each pump is constructed to deliver 9150 cubic feet—255 tons—a minute, or a total for the eight machines of 2040 tons. The pumps are placed one on each side of the engines, the pump shafts forming prolongations of the crank shaft, and being connected to the latter by disc couplings. The pump shafts are of steel, $8\frac{1}{2}$ inches diameter, and are provided with bearings beyond the pump casings. The pumps have discs 5 feet 9 inches diameter, with delivery pipes 54 inches diameter, and double-suction pipes, in area jointly equal to the delivery pipes. The casing of each pump is made in a single casting, 15 feet diameter. The engines have cylinders $27\frac{3}{4}$ inches and $46\frac{5}{8}$ inches diameter, the stroke being 2 feet 3 inches; both cylinders are jacketed. For some years after the starting of these machines the low-pressure cylinders of each engine exhausted into a pair of surface condensers, placed on the discharge pipes. These condensers were cylindrical chambers traversed by a number of 3-inch tubes, connected with the pump casing and discharge pipe. It was suggested that the efficiency of the pumps was interfered with by the presence of these surface condensers in the delivery pipes. They were removed, and condensation by injection, with auxiliary air-pumping engines, substituted. The alteration was, however, a doubtful improvement, the difference in efficiency was not very observable, and the auxiliary engines involve extra attention. Steam is generated by two groups of boilers, each consisting of five, of a compound, double-flued type, with Galloway tubes and horizontal marine tubes. At the official trial made in May 1875 the consumption of fuel was $2\frac{1}{2}$ lb. per indicated horse-power per hour, or 4 lb. per horse-power of water lifted—doubtless the best result obtained on drainage works up to that time. All these pumps worked continuously day and night from the 10th October, 1878, to 31st May, 1879, in a satisfactory manner. Since the latter date the seasons have been drier and the work lighter. These engines and pumps are at the present time in perfect order, the cost for repair having been very light. In April 1887 Messrs. J. and H. Gwynne received a certificate from the engineer-in-chief of the pumping station, Sig. Ardizzoni, stating that, notwithstanding the excessive work they had undergone since their erection, the pumps had never wanted the slightest repair, and that the machinery was then in perfect working condition, and splendid results had been obtained from it. About the same time that these pumps were put up by Messrs. Gwynne, four scoop wheels were fixed by a Dutch firm for draining the Marrozzo Marshes on the other side of the river Volano. In accordance with the recommendation of a commission of engineers these were afterwards taken down and replaced by Messrs. J. and H. Gwynne with two of their centrifugal pumps calculated to discharge, each, 9951 cubic feet (277 tons) a minute. The new pumps are worked by the existing engines, and in addition a supplementary compound engine and pump, to discharge 70 tons a minute, is added. Although the large pumps measure 17 feet over their cases, their weight is hardly one-fourth of that of the four wheels which they replaced. The old foundations were utilised to a large extent, and the suction pipes dip into the old wheel pits. The pumps, since they have been put to work, have acted very successfully, keeping the water down with great ease during the winter, only one pump being required. It is considered that one of the large pumps does as much work as was done by the four wheels combined.

A plan showing the general arrangement of these pumps and an elevation of the building will be found on Plate 7.

Fos, Bouches-du-Rhône, South of France.—In this district large reclamation works were carried out in 1884–85. Pumping was required, and Messrs. J. and H. Gwynne were commissioned to erect at Fos a pair of their "Invincible" compound direct-acting centrifugal pumping engines, each pump to raise 60 to 70 tons per minute;

and at another pumping station (Gallejon) some miles distant, a third "Invincible" engine was provided, to raise 90 to 100 tons per minute. The lift was low (1·0 metre to 1·80 metre), and experience has shown that with such lifts a low efficiency was to be expected. The makers, nevertheless, guaranteed that for the smaller machines the steam used would not exceed ·0718 kilogs. for each cubic metre of water raised one metre, and that the steam per indicated horse-power per hour would be from 20 to 24 lb. English. For the larger machine the figures were ·07 kilog. and 20 to 22 lb. English. Very carefully conducted trials were made with the smaller pumps, by Mr. A. C. J. Vreedenberg, a Dutch engineer, when the following results were obtained from No. 2 engine, with a mean lift of 1·379 metres (4·52 feet):—Water raised, each pump per minute, 65·7 tons; horse-power in water lifted, 20·18; horse-power indicated, 37·0; efficiency, ·54; coal consumption per W.H.P. per hour, 4·45 lb.—2·033 kilogs. Owing to a slight defect in the machinery, discovered after the trial of No. 1 engine, the results obtained were not quite as good as those from the other engine; but this having been remedied, there was found to be very little difference in the working of the two sets of machines. Given on terms of the guarantee, the result was ·0627 kilog. of steam per cubic metre of water raised 1 metre, and 20·66 lb. English per indicated horse-power per hour. The boilers were of French design and construction, their evaporative efficiency being not very high, 8·33 lb. of water per pound of good coal. The coal per indicated horse-power was 2·47 lb. This was subsequently confirmed by trials of eight hours on each pump, conducted by Mr. Dornés, engineer of the reclamation, who obtained with one pump the following results :—lift 65·75 inches ; mean delivery, 64·7 tons; efficiency of whole machine, ·579; steam per indicated horse-power, 20·32 lb.; coal per indicated horse-power, 2·44 lb. per hour; coal per water horse-power, 4·21 lb. per hour. The result from the other pump was practically the same.

On the larger pump at Gallejon some interesting experiments were made by French engineers with progressive speeds and lifts. The lifts varied from $3\frac{1}{4}$ feet to $6\frac{1}{2}$ feet. Three runs each of thirty minutes were made with each lift, and with a different number of revolutions for each half-hour. The results showed that an efficiency of ·50 was obtainable on a lift of 3·28 feet; on the other lifts the average efficiencies were for $4\frac{1}{3}$ feet, ·56; for 5 feet, ·607; for 6 feet, ·66; and for $6\frac{1}{2}$ feet, ·698. All efficiencies represent, as in the experiments with smaller pumps, the ratio which the water horse-power bears to the indicated horse-power. The details of these trials are given in the following table. These short experiments may not give results so strictly accurate as more lengthened trials, but they are consistent in agreeing with those obtained from the smaller pumps, and it is evident that in these machines remarkably high efficiencies were obtained, considering the small horse-power and the low lifts. The difficulty in preventing waste of energy while raising a large volume of water rapidly through a small height is obvious enough, pump and engine friction making a larger fraction of the total power uselessly expended when the lift and water horse-power are small. It is important to observe in these experiments the steady increase of efficiency with increase of lift.

Trials of the Gallejon Pumps.—Table of discharge and efficiency corresponding to different heights of lift and to variable speeds.

Height of Lift in Metres.	Revolutions.	Discharge in Litres per Second.	Efficiency between work shown on pistons and effective work in water raised.
1·000 (0·900 to 1·100)	65·4	458	0·325
	71·5	716	0·428
	86·5	1280	0·446
	101·3	1634	0.504
1·322 (1·300 to 1·350)	90·6	1241	0·603
	107·7	1748	0·568
	117·5	1990	0·511
1·540 (1·520 to 1·560)	93·0	1070	0·622
	103·0	1461	0·618
	116·0	1884	0·581
1·800 (1·760 to 1·840)	94·0	955	0·658
	106·5	1448	0·797
	120·0	1892	0·634
2·000	107·0	1314	0·718
	119·0	1738	0·678

FONDI, SOUTHERN ITALY.—In the year 1882 the Provincial Board of Public Works of Caserta entered into a contract with Messrs. Guppy and Co., of Naples, for two complete sets of steam engines with centrifugal pumps, which were guaranteed to raise 20,500 gallons (equal to 91·51 tons) of water per minute to a height of 7 feet 8 inches, for the purpose of draining part of the marshes near Fondi, for reclaiming the land for cultivation, and at the same time rendering the neighbourhood more healthy. This large extent of constantly submerged land was useless for cultivation, and the stagnant water made the whole district a most unhealthy one, as it produced pestilential exhalations during the summer months, which infected and poisoned the atmosphere for miles around. On this account the peasants could not reside on the spot, but lived in villages built on high ground a considerable distance from the marshes, and came down to their work every morning, returning home in the evening before sunset. Even with these precautions it was scarcely possible to escape the effects of the bad air, and the people who frequented these districts were sallow, subject to low intermittent fevers, and not in a really fit state of health to perform hard work. The value of the land for these reasons was naturally much depreciated, and in many cases the owners allowed their property to be confiscated rather than submit to pay taxes on ground which brought them no return.

The marshes which are now drained are situated between a range of hills on the north, and the small Lake of Fondi on the south, and are formed by a considerable depression of ground, which permitted the rain-water to collect. The lake derives its supply of water from springs at the base of these hills, and the water is brought down by the Canale d'Acqua Chiara, the overflow discharging itself into the sea on the opposite side. The mode of draining these marshes has been to cut large canals parallel to each other, intersecting the space between by small canals that collect and bring the water down to the pumping station. These large canals vary in width from 6 to 15 feet, their total length being upwards of twelve miles; the small canals are about 4 feet wide and over six miles long altogether. Several of these canals extend down to the sea, and are there provided with sluices, which are closed when the tide rises. The average annual rainfall is 32 inches; but as a large volume of water collects from the surrounding country, it has been found from experience that it requires at least six days to pump up the rain-water that falls in twenty-four hours. The pumps generally commence working after the first heavy rains in October, and continue running, as required, throughout the winter up to the end of April. On starting the pumps in the autumn, when the land requires to be tilled, both engines and pumps have to work for the first fifteen days, twenty-four hours per day, until the water is reduced to a certain level; for the next fifteen days they work twelve hours a day, and after this one engine and pump is generally sufficient to keep the land dry, except in a wet season, when, if necessary, both engines are employed to pump away the extra rainfall. All the water pumped up is discharged into the Canale d'Acqua Chiara, and flows through the Lake of Fondi into the sea. It may be stated that the quantity of water lifted by the pumps has been ascertained to be about 27,000 gallons (equal to 112·5 tons) per minute, which is equal to about $\frac{1}{4}$ of an inch of rainfall in 24 hours.

The area of land drained is about 12,000 acres, of which at least 10,000 acres are at present kept dry and under cultivation. The land reclaimed is a rich alluvial soil, mixed with clay, admirably suited for raising all kinds of cereal crops, especially Indian corn. Since it has been drained and cultivated there is decided improvement, splendid crops are being raised, and the proprietors, far from demurring to pay the land-tax, now willingly submit to an extra tax for the drainage of their property, which lets freely at 5*l*. per acre.

The house erected to contain the pumping machinery is situated close to the Canale d'Acqua Chiara. Owing to the nature of the soil, it was necessary to drive in piles on which broad foundations were laid and walls built. On each side of this building are the lateral canals that bring the water from the marshes to the suction pipes of the pumps, whence it is lifted and discharged into the large reservoir constructed in masonry, which is in communication with the Canale d'Acqua Chiara, so that the water flows away direct to the lake. The reservoir has sluices that can be closed should the water in the Canale d'Acqua Chiara rise higher than the level of the water in the reservoir.

There are two horizontal boilers, each 18 feet long and 4 feet 11 inches in diameter, with an internal flue 2 feet $7\frac{1}{2}$ inches in diameter, in which are placed five conical tubes; at the back end of the boiler there are thirty-one iron tubes. Each boiler has a heating surface of $376\frac{1}{2}$ square feet, and a fire-grate area of 12 square feet. The working pressure is 70 lb. per square inch; and as a single boiler is sufficient for supplying the steam to both engines, the second is kept in reserve.

The two horizontal steam engines that drive the centrifugal pumps have one cylinder each of $13\frac{3}{8}$ inches in diameter, with $15\frac{3}{4}$ inches in length of stroke, fitted with variable expansion valves, cutting off steam at 0·175 of the stroke. The condensers and air-pumps are placed behind the cylinders, being bolted to the same bed-plate, and are worked direct by the piston-rods of the cylinders. In case the suction pipes of the air-pumps should get choked with weeds, an extra injection cock is provided, connected with the casing of the centrifugal pump, so that the air-pump may still obtain its supply of water for condensing. This cock can also be used to form a vacuum in the cases of the centrifugal pumps, should the ejectors get out of order. The small donkey engine for feeding the boilers draws water from the condenser when the engines are working or from the canal when they are at rest.

The two large centrifugal pumps are coupled to the main shaft of each engine. The outer casing of these pumps is 7 feet 7 inches in diameter, and cast in two pieces, the lower part being fixed to one side of the bed-plate of the engine, and has a large flange on the upper side, on to which the upper half is bolted. The lateral suction pipes are also bolted to the sides of this casing by flanges, which admit of their being easily dismounted to examine the disc inside, remove weeds, &c. The suction pipes have no foot valves, and the discharge pipe a hinged valve, which remains closed when the pumps are not at work. The discs are 3 feet $11\frac{1}{2}$ inches diameter, with a central web having six long and six short blades on each side. The spindle of the pump is of Bessemer steel, with a brass casing and lignum-vitæ bearings. Each pump is supplied with an ejector for forming the vacuum in the suction pipes before starting.

The following experiments were made in December 1882 in the presence of the Italian Government engineers appointed to inspect the machinery, and to ascertain if it fulfilled the conditions of the contract. At the two trials both engines were set in motion with steam supplied from one boiler, and were kept running for twelve hours. On the first occasion the pumps ran at 130 revolutions per minute, the pressure of steam in the boilers was 70 lb. per square inch, cut off at 0·175 of stroke, the vacuum was $24\frac{1}{2}$ inches, and 24,700 gallons of water were raised per minute, the lift being 6 feet $6\frac{3}{4}$ inches, and the consumption of coal $2\frac{3}{4}$ cwts. per hour. On the next trial the experiment was continued with a lift of 5 feet 8 inches, when, with the same number of revolutions, the pumps raised 27,300 gallons of water per minute, with a consumption of $2\frac{1}{2}$ cwts. of coal per hour. The horse-power in water raised was on an average $47\frac{1}{4}$ per cent. of the indicated horse-power, with a mean consumption of 6·12 lb. of coal per horse-power per hour.

[Extracted from a paper by Mr. T. R. Guppy in the 'Minutes of the Proceedings' of the Institution of Civil Engineers.]

LAKE HAARLEM, HOLLAND.—This tract of land was originally a large fresh-water lake, which it was supposed had been caused by inundations in 1591 and 1647, previous to which time it had been an inhabited district with three villages. In shape it is an irregular oblong, the length from north to south being 14½ miles, and the greatest width eight miles. The total area contained 56,609 acres of lowlands and meres, and formed the "boezem" or collecting basin for the surrounding lands, being a portion of the great drainage district of Rijnland. The surface of the water in this boezem was maintained at its lowest level by natural drainage, through sluices emptying into the North Sea—one at Katwijk and the others into the Y at Spaarndam, and at Halfweg. Schemes for the drainage of this lake date back two and a half centuries. In 1643 Jan Adriansz—surnamed Leeghwater—a millwright, published a detail plan for the drainage, which passed through thirteen editions, the latest appearing in 1838. In 1836 very severe storms occurred which drove the water of the lake upon Amsterdam, and up to Leyden, submerging part of the city and inundating 100,000 acres of polders. These disasters finally decided the Dutch States-General in decreeing the reclamation of the lake, and in 1839 a vote, amounting to over three-quarters of a million of money, for the purpose was passed. It was not, however, until nearly ten years afterwards that operations were actually commenced. The first work was to surround the lake with a dyke or bank to shut off the water from the adjoining polder. Parallel with the bank a canal was cut called the Ringvart. The dyke and canal were 37 miles long; the top of the dyke was 7½ feet above A.P.—or 9·63 feet above ordinary high-water in the North Sea—and the bottom of the canal, 19½ feet below A.P. The canal was 140 feet wide, having a depth of 10 feet for a width of 95 feet. and navigable for vessels. A road was made between the canal and the dyke. The canal had slopes of two to one, and, with the cess, occupied an area of 654½ acres. The dyke was made of peat, and occupied, with its slopes, 1013½ acres.

A commission was appointed to determine as to the most suitable machines for raising the water from the lake, and for afterwards keeping it dry. The use of windmills, driving scoop wheels, or Archimedean screw pumps, was strongly advocated, while the advantage of steam was also pressed on the attention of the Commissioners. It was found after fully investigating all the proposals that the estimated cost of draining the lake by wind-power would be over 300,000*l.*, and that the maintenance of the 114 windmills required would amount to over 6000*l.* a year. The cost of draining by steam-power was estimated at 100,000*l.*, and the annual expenditure after the lake was dry at 4500*l.* Some of the Commissioners came over to England to inspect the steam pumps in the mining districts of Cornwall and elsewhere, and as the result of their inspection recommended a design submitted to them by Messrs. Gibbs and Deane, which was finally adopted. The dimensions of the steam engines and pumps as set out in this design were larger than any that had previously been constructed, and the whole scheme was so novel, and differed so much from anything that had ever been attempted before in Holland, that considerable anxiety was felt by the Commission in incurring so large an outlay. The agreement with Messrs. Gibbs and Deane stipulated that they were to receive a premium of 3000 guilders, whether the machinery succeeded or not; if successful, to have 9100 guilders in addition—making about 1000*l.*—and 200 guilders for each million pounds in excess of the stipulated 75,000,000 lb. of water raised 1 foot high with 94 lb. of best Welsh coal.

Observations on the rainfall of Holland, extending over a period of ninety-eight years, had shown that the greatest depth of rain in any one month was 6·524 inches more than the evaporation for the same period; 1·47 inches were allowed for infiltration, giving 8 inches to be lifted in one month. The level of the lowest land was 14 feet below A.P., and the water in the drains, after the lake was reclaimed, was settled to be 15½ feet below A.P. The lift into the Ringvart would therefore be, when the pumping was completed, 15½ feet. The lake at 13 feet deep contained 780,000,000 tons of water, which had to be lifted an average height of 16½ feet. The total rainfall and infiltration was estimated at 40,000,000 tons during the works and 60,000,000 tons afterwards. It was calculated that to lift this quantity and afterwards to keep the polder dry, three engines of 350 horse-power each would be required, and it was determined to erect these at the extremities of the lake. These engines were subsequently named the Leeghwater, at the south, near Kaag; the Lynden, at the north; and the Cruquis, near the junction of the canal with the Spaarn. It was estimated that, with no delay from accidents, the lake could be laid dry in fourteen months, allowing 250 working days in a year. From a variety of causes, and the access of water from infiltration beyond what was anticipated, the time actually occupied was thirty-nine months, and the quantity raised 900,000,000 tons. The time the pumps were actually at work was nineteen and a half months, frequent delays occurring from the valves becoming choked with silt and other causes. In the winter season the rainfall, with the absence of evaporation, gained upon the power of the engines. The general average lowering of the surface was at the rate of about 4 inches per month, every inch in depth giving 4,000,000 tons.

The dyke and canal were finally completed in 1848, and pumping began with the Leeghwater, and with the other engines in the spring of the following year. The lake was laid dry in 1852. The works for fitting the bed of the lake for cultivation consisted in making main canals 80 feet wide along the centre from north to south and from east to west, terminating at their respective ends at the three pumping stations. Four smaller canals parallel with the others were made lengthways, and six across the lake. The subdivisions contained fifty acres each. The length of the large canals was 18·63 miles; of the smaller, 93·15 miles. The total length of the canals and ditches was 750 miles. There were also made 122 miles of roads and sixty-five bridges. The total area of the polder to the encircling canal is 41,648 acres, of which 3011 acres, or about 6¾ per cent., are occupied by roads and main waterways. The estimated expenditure at the commencement of the operations was 687,500*l.* The total amount actually expended was 781,500*l.*—at the rate of 16·57*l.* per acre of available land, out of which the dyke and canal cost 161,527*l.*; the engines, boilers, and pumps, 71,216*l.*; and buildings for same, 50,615*l.* Taking the power in water lifted at 1037 horse-power, this is equal to 48·23*l.* for buildings; 68·66*l.* for machinery; together, 116·89*l.* per horse-power. The net cost after sale of the lands, exclusive of interest and commission, was 86,042*l.*, for which 41,648 acres were added to the taxable resources of the country, and provision made for the maintenance of a large population, now amounting to over 12,000. The first public sale of the land took place in 1853. The prices realised ranged from 25*l.*, the average over the whole polder being about 16*l.* an acre.

The average rainfall on the polder for the ten years ending 1872 was 31·267 inches. The time the pumps were working was 5584¾ hours, divided as follows:—

	Rainfall. inches.	Hours pumps working.
First four months	7·472	2254¼
Second ,,	10·503	398½
Third ,,	13·292	2932

The average annual consumption of coal was, during the same period, 2690 tons. The rainfall for the ten years—1877–86—averaged 32 inches, the maximum being 39·13 inches in 1877; and the minimum, 26·69 inches in 1884. The average number of hours the pumps were working, during the same period, was 6823 hours; equal to ninety-four days of twenty-four hours for each station, the average quantity of water pumped being 96,091,600 tons. The water is kept at an average level of at least 3 feet below the surface of the arable land, and about 2 feet for the grass land.

The Leeghwater was the first machine erected; the three sets of machinery are almost similar in design and dimensions. The engines are beam engines of the Cornish type, single-acting, condensing, and working expansively, giving motion to a series of pumps, working at a single lift arranged concentrically round the engine. The water is delivered by the buckets of the pumps on to a spilling-floor A, Plate 8, Fig. 13, at either side of which are self-acting doors B, which open out to a short channel leading to the main canal. These doors open as soon as the water on the floor rises above the level of that in the canal, and close as soon as the pumps cease. There are valves which, when opened, leave the spilling-floor dry, so that the buckets and other parts of the pumps can be drawn out and laid on the floor for repair. The general arrangement of these engines will be seen from the drawing, Fig. 14. The foundation for the machinery and buildings consists of 1400 piles driven to a depth of 40 feet into a stratum of hard sand. On these a platform was laid 21 feet below the surface of the lake, and upon this a brick well was built in which the pumps were fixed. The engine was placed in the centre, with the pumps C, eleven in number, ranging round three sides in the segment of a circle, the boilers being placed at the back. The engine is of peculiar construction, having two cylinders, D, E, one within the other, united at the bottom, and having a clear space of $1\frac{1}{2}$ inches between them at the top under the cover, which is common to both. The outer or annular cylinder, E, is 12 feet in diameter, and the inner, D, 7 feet. The pistons are connected to the rocking-beams by one main piston-rod attached to the smaller piston, 12 inches in diameter, and four small rods attached to the annular piston, each $4\frac{1}{2}$ inches in diameter, and having a large crosshead with a circular body 9 feet 6 inches in diameter, and formed to receive the ends of the balance-beams of the pumps. When the pistons are at the bottom of the cylinder, steam is admitted at a pressure of from 40 lb. to 45 lb. beneath the interior piston, which is then raised, carrying with it the annular piston, crosshead, and a weight of about 30 tons of iron, with which it is loaded, the total dead-weight lifted being about 100 tons. Steam is cut off at $\frac{8}{10}$ths of the stroke, at the end of which an equilibrium valve is opened and the steam admitted to both sides of the piston, which is held in its place for a short interval by two hydraulic rams, one on each side of the cylinder, in order to enable the pump valves to adjust themselves; the equilibrium valve is then closed, the eduction valve opened, and the steam passes through the large cylinder to the condenser, the piston descending by gravity, drawing down the balance-beam of the pumps, and lifting the water on to the upper floor of the well. The various valves for the admission of steam to the cylinder, the equilibrium valve, and to the condenser, and for the cataract, are opened and closed by tappets on rods, which strike levers actuating the valves.

The pumps at this station are eleven in number, and each of them 63 inches in diameter; at the two other stations there are only eight pumps, each 73 inches in diameter. The pumps are attached to cast-iron balance-beams turning upon a centre in the wall of the engine-house, the other end of the beam being connected with the crosshead of the engine. Each pump-rod is of wrought iron, 3 inches in diameter and 16 feet long. The steam and pump pistons have a stroke of 10 feet. The eleven pumps make ten strokes a minute, and raise each 6 tons per stroke—equal to 660 tons per minute. The weight of the several parts is as follows:—beams, 9·82 tons each; the cylinder, 24·2 tons; crossheads, 18·8 tons; pump cylinders, 6·82 tons each; and buckets, 3 tons.

The first trial of the Leeghwater was made in September 1845, and was in full working order in the following November. The engine was found at the trial to do a duty equal to raising 75,000,000 lb. one foot high by the consumption of 94 lb. of good Welsh coal, and exerting a net effective force of 350 horse-power with a lift of 13 feet. There are five boilers, each 30 feet long by 6 feet in diameter, with a single flue 4 feet in diameter. The Cruquis has eight pumps 6 feet in diameter with 10 feet stroke, lifts 8 tons per stroke, making ten strokes a minute, or a total for all the pumps of 640 tons per minute, the average lift being 15·58 feet. The Lynden machinery is similar to the Cruquis—it works generally during the winter, using only seven pumps, making seven strokes a minute;

L

delivery, 7 tons each stroke, with an average lift of 15 feet; the steam pressure in the boiler being 40 lb., and cut-off in the cylinder at half the stroke.

The coefficient of useful effect of these pumps is stated by Sig. Cuppari to be higher than any other machines in Holland, the ratio of effective to calculated discharge for the Cruquis being 89 per cent., and for the Leeghwater rather less. The ratio of horse-power in water lifted to that indicated of the Cruquis is $\frac{340}{560} \frac{\text{W.H.P.}}{\text{I.H.P.}} = 60 \cdot 60$ per cent. The average consumption of coal over a long period is 6·82 lb. per W.H.P. per hour. At trials made by Mr. Elink Sterk, the engineer of the Haarlemermeer, it was found that with the engine making three strokes a minute the consumption of coal was 5·44 lb. per I.H.P., and 7·80 lb. per W.H.P.; the ratio of efficiency of the machinery being ·698. With seven strokes a minute 4·94 lb. of coal were used per I.H.P., and 6·74 lb. per W.H.P. The ratio of $\frac{\text{W.H.P.}}{\text{I.H.P.}} \frac{361}{492}$ being ·734.

To compensate the district of Rijnland for the loss of the large area of polders taken from it by the drainage of Lake Haarlem, and to effect the maintenance of the water in the Ringvart or surrounding canal at a uniform level, pumping stations were erected at Halfweg, half-way between Haarlem and Amsterdam, to lift the water into the Y, when it rose above the height at which it would flow away by gravitation; at Spaarndam, near Amsterdam, to lift it into the Spaarn; and at Katwijk, to lift it into the North Sea. A pumping station was also erected at Gouda, to regulate the water in the Gouda Canal, and to discharge the flood water into the Yssel. The table on p. 147 gives the principal dimensions and particulars of the engines and wheels.

Halfweg.—The six scoop wheels * at this station are ranged in two sets of three each on either side of the engine-house. The wheels have a combined width of 39·36 feet. Each set of these wheels is fixed on a cast-iron axle, but is so arranged that the couplings can be disconnected and only part of the wheels worked when the lift is high. The framing for each of the wheels consists of three heavy

---

* Illustrations of the scoop wheels at Katwijk and Halfweg, and also of other drainage machinery, will be found in the atlas of plates of 'Stoombemaling van Polders en Boezems,' by A. Huet, C.E., published at The Hague.

Pumping Stations.

	Halfweg.	Katwijk.	Spaarndam.	Gouda.
Date of erection	1852	1880	1845	1857
Number of wheels ..	6	6	10	6
Dia. of wheels in feet ..	21·64	29·50	17·05	26·00
Width in feet	6·56	8·00	..	5·75
Total width	39·36	48·00	68·42	34·50
Number of scoops ..	24	..	..	12
Internal diameter, feet	11·84	..	..	..
Dia. of circle to which scoops are tangents in feet	5·70	10·50	..	11·16
Number of revs. per min.	4½	4½	8	4
Velocity of periphery in feet per second ..	5·00	6·17	..	5·44
Weight of each wheel in tons	15	43	..	..
Average dip of wheels in feet	4·60	5·66	..	4·16
Lift in feet	1·64 to 2·62	0 to 7·00	..	4·0 to 6·0
Total discharge in tons per minute	1037	2000	1869	1053
Description of engine ..	single-cylinder horizontal condensing	2 compound horizontal condensing	2-cylinder horizontal condensing	horizontal condensing
W.H.P.	123	370	280	240
Ratio of speed of engine to wheel	13·5 to 6·00	36 to 4	3 to 1	4 to 1
Ratio of W.H.P. to I.H.P. per cent. ..	..	{ 33 to 70, mean 50 }	..	81·97
Days of 24 hours working 1 year	50, average 3 years, 1853-56	60	..	..
Pressure of steam in lbs. per square inch	45	80	..	65
Cylinder dia. in feet ..	3·36	2·00 and 3·00	3·02	3·00
Length of stroke	8·00	4·25	5·16	5·25
Number of boilers ..	3	8	4	3
Size of boilers in feet ..	28·0 × 5·50	32·66 × 7·31	37·72 × 5·40	..
Coal used in lbs. per W.H.P. per hour ..	5·50	7·71	11·00	..
Cost of machinery in £	4845	14,200	..	..
Do. per W.H.P.	39·39	23·66	..	..
Do. buildings in £ ..	7896	15,100	..	..
Per W.H.P.	64·20	25·16	..	..
Do. total do.	103·59	48·82	..	..

rings of cast iron, with spokes connecting the rings to an iron nave. These rings are cast in two segments and bolted together, and to them are attached cast-iron start-posts, and on these twenty-four flat wooden scoops are attached. The consumption of coal has been found to vary very considerably as the lift is altered, owing to the large proportional absorption of power at the low lifts required simply to move the machines, the following being given as the result of trials :—

Lift in feet.	Consumption of coal in lb. per horse-power per hour in water lifted.
0·45	50·0
0·66 to 1·00	14·20
1·00 to 1·33	11·00
1·33 to 1·66	8·80
2·00 to 2·28	5·50

The discharge of these wheels varies from 154 tons per revolution when the water is low in the canal, to 230 tons when it is at its highest. These wheels ran 3623 hours during the emptying of Lake Haarlem, the average lift being 20 inches. The average power exerted in water lifted was 92 horse-power, and consumption of coal at the rate of 9 lb. per horse-power per hour. The total quantity of water raised was 202,765,406 tons. The time the engines ran for the three years after the lake was dried—May 1st, 1853, to July 1st, 1856—was 3675 hours, equal to an average of fifty-one days a year.

Katwijk.—The combined machinery at Katwijk is the largest instalment of pumping machinery erected in Holland or England. The arrangement and construction of the wheels is similar to that at Halfweg. The wheels are capable of lifting 2000 tons per minute to a height of 4 feet, or 1200 tons to a height of 7 feet. On the discharge side the floor of the raceway immediately in front of the wheel is made movable and hinged at the outer end, so that it rises and falls automatically according to the height of the water, and so makes a movable breast and prevents the back current on to the wheel when it is working. The height to which the wheels have to lift the water varies daily with the tidal condition of the sea from a few inches to 7 feet. The variation on the inside is small, never altering more than about 18 inches. The average time of working is about sixty days of twenty-four hours in the year.

Gouda.—The six wheels here are all ranged in two sets on one axle. When the station was first erected the wheels had flat scoops, and were driven by a high-pressure condensing horizontal engine, having an effective power of 111 W.H.P., made by the Atlas Company at Amsterdam. The wheels were each 5·75 feet wide and 24·27 feet in diameter. The consumption of coal with this machinery was at the rate of about 6 lb. per W.H.P. per hour. The cost was 7863*l*. for buildings and 10,666*l*. for machinery, equal to about 70·84*l*. for

Pumping Stations. 149

the former and 96*l*. for the latter, or a total of 166·84*l*. per horse-power of water lifted. In 1872 these wheels were changed for wheel pumps—*Pompraderen*—having a width of 5·25 feet, and diameter 25·84 feet. The drum was 19·41 feet in diameter. These wheels made from 3·63 to 4·30 revolutions a minute. The number of buckets was originally six, this number being subsequently increased to twelve. At one time buckets were tried, having a curved form with the concavity towards the inner water, and others with the concavity towards the outer water. There was not found in working to be much difference in the result between the two systems, but, if anything, the latter gave the best results. The loss by slipping and imperfect filling of the pump wheels was found to amount to as much as 22 per cent. of the theoretical discharge. These pump wheels were recently altered to ordinary wheels having flat scoops, and new engines of the horizontal condensing type have been put up by Friedrich-Wilhelms-Hütte, of Mulheim. These engines have a stroke of 5·25 feet; cylinder diameter, 3·00 feet; I.H.P., 147 each. Steam is used at a pressure of 60 lb., and is cut off at half the stroke. The engines make 16 revolutions per minute to 4 of the wheel. With four wheels at work the discharge is 152·8 tons, lifted to a height of 6 feet, equal to 240 water horse-power. The indicated horse-power of the two engines being 294, this gives an efficiency of 81·97 per cent. The scoops are tangents to a circle having a diameter of 11·16 feet. The outer diameter of the wheel is 26 feet; the inner diameter and length of scoops, 4·23 feet. The scoops are twelve in number, made of wood; the framing of the wheel being iron. With the scoop fully immersed, the discharge at each wheel is 43·88 tons per revolution, or 175·52 tons per minute, and for the six wheels 1053·12 tons. With an immersion of 3 feet the discharge is reduced to 33·22 tons per revolution. The two engines use at the rate of half a ton of coal per hour, which, with an effective power of 240 water horse-power, is equal to 4·67 lb. per hour.

BULLEWIJKER POLDER, HOLLAND.—This machinery was put up in 1881 by Messrs. J. and H. Gwynne; it is of a similar type to that erected by this firm elsewhere, and is capable of raising 60 tons a minute. The lift is 14 feet—4·67 metres; maximum discharge, 2508 cubic feet per minute—1·184 cm. per second; coal consumed per W.H.P. per hour, 5·22 lb.—2·37 kilogs.; ratio of water to indicated horse-power, 58 per cent. The trials from which these results were obtained were made by Mr. Elink Sterk, the engineer of the Haarlemermeer drainage. This pump has very long suction and delivery pipes, doubtless reducing its efficiency to a small extent.

BIJLMERMEER, HOLLAND.—The first compound centrifugal pumping engine put down in Holland was supplied by Messrs. J. and H. Gwynne for this drainage in 1883. When tested, the pumps raised 70 tons per minute 14 feet high; the ratio of water to indicated horse-power was ·613; and the coal used, German, was 4·67 lb.—2·12 metres—per W.H.P. per hour. Although German coal is inferior to good English steam coal, the consumption would not have been so high as stated had the boilers been more perfectly proportioned to the requirements of the engines. Two are provided, one is rather too small to give steam enough, while both are quite too large.

ZUIDPLAS POLDER, NEAR MODERICHT, HOLLAND.—This is a very low polder, the peat having been used largely for fuel. The lift from the surface of the water to the mean level of the river Yssel, into which the water is discharged, is about 22 feet. The area of the polder is 11,050 acres, of which 682 acres are occupied by the Ringdyke, leaving 10,368 acres. At the beginning of the present century windmills driving scoop wheels were erected to keep down the level of the water. In 1825 the Dutch Government determined to thoroughly drain the lake. For this purpose it was divided into two parts, the lower level being separated from the upper by a surrounding canal or Ringvart, into which the water from the lowest part was lifted, and from which it was raised with the drainage of the rest of the polder into a collecting basin, which discharged into the river Yssel by sluices. The lower level was drained by fifteen windmills, eight driving Archimedean screws, and seven scoop wheels. For the lift from the upper level there were fifteen windmills, ten of which drove screws and five wheels. There were also erected in 1838 two steam engines driving screws. The windmills were in operation until 1873, the annual cost of maintaining them being at the rate of 60*l.* each. The Archimedean screws, erected in 1838, were 5·83 feet in diameter, the axle being 1·70 feet, and the blades 0·12 foot thick. The axle was laid at an inclination of 30°. The screws made 47 revolutions a minute, and the engines 19·86. The engines were direct double-action, with cylinders 1·66 feet diameter, and having 7 feet stroke. The spur wheel fixed on the crank shaft, geared into a bevel wheel on the axle of the screws. With a lift of 6·16 feet, the discharge was 1·17 tons per revolution, equal to

57·46 tons per minute. The horse-power of the engine in water lifted was 24 water horse power. The consumption of coal was at the rate of 21·34 lb. per water horse-power per hour.

There are now two pairs of pumping stations, each having installations consisting of scoop wheels and centrifugal pumps. The scoop wheels at present in use are the largest in diameter in use in Holland, one being for the lower lift and the other for raising the water from the Ringvart to the collecting basin. The wheels are 32·80 feet in diameter and 4 feet wide. The lift is 11·80 feet for the lower, and 8·20 feet for the upper wheel. They have thirty-two curved iron scoops, the curve being concave to the internal water, and make 4½ revolutions a minute. The velocity at the periphery is 7·22 feet per second; the weight of one wheel with its axle is 21 tons. The author has seen these wheels at work, his impression being that their performance is not satisfactory, the scoops not entering or leaving quietly, or with the best effect, the water being too much dashed about. The wheels are not provided with shuttles to regulate the supply of water to the scoops. These wheels work in an uncovered masonry raceway outside the engine-house, and are driven by single-cylinder condensing horizontal engines. The wheels discharge 110·5 tons per minute to a height of 11·75 feet; the horse-power of the engines in water lifted is eighty-nine, and the consumption of coal at the rate of 7 lb. per water horse-power per hour. The two steam engines and wheels cost 5000*l*., or at the rate of 28·03*l*. per water horse-power.

The centrifugal pumps were erected in 1876 by Messrs. Gwynne and Co., London. They are of the direct-acting type, having horizontal spindles, and are fixed in pairs in the two engine-houses, one for the lower and the other for the upper levels. The suction and delivery pipes are 3 feet in diameter, and the latter is carried by a bend below the lowest water-level in the basin into which it discharges. Each station has two pumps driven by separate non-compound direct-acting engines, fitted with variable expansion gear, and having steam-jacketed cylinders, 24 inches diameter, with 20 inches stroke; the engines and pumps making 100 strokes per minute at the full lift of 13 feet. A separate air pump is provided for charging the pumps at starting. Each pump is capable of raising 71 tons per minute. The diameter of the pump disc is 6 feet, and the width 3 feet. Steam is supplied by four Lancashire boilers, with 24 Galloway tubes 25 feet long by 6·5 feet wide, at the lower station; and three boilers at the upper. The working pressure of steam is 75 lb. The lift varies from 5 feet to 13 feet. These machines give off in water lifted from 40 to 50 per cent. of the indicated power, the average being 40. The consumption of coal is at the rate of 7·67 lb. per water horse-power per hour. The four centrifugal pumps, with the engines and boilers, cost 11,833*l*. Taking the power in water lifted for such as 62·64, this is at the rate of 47·22*l*. per water horse-power. The pumps work under some disadvantage as compared with the wheels, the work of the latter being more regular and constant, while that of the pumps frequently varies both with regard to the lift and time of working. At a trial of one of these pumps made in 1877, the consumption of coal was found to be at the rate of 5·95 lb.—2·7 kilogs.—per horse-power of water lifted. The average lift was 12·73 feet; the number of revolutions, 105; discharge, 87·17 tons per minute.

BEEMSTER POLDER, NORTH HOLLAND.—A pair of direct-acting engines and centrifugal pumps were fixed for the drainage of this polder by Messrs. Gwynne and Co. in 1878. Each of these engines is stated to be capable of raising 100 tons of water a minute to a height of 17 feet. The cylinders of the engines are steam jacketed, 27 inches in diameter, with 20 inches stroke. The engines work at the rate of 100 revolutions a minute. The steam in the boiler is supplied at a pressure of 75 lb. per square inch, and is cut off at one-eighth of the stroke by adjustable expansion valves. Four smaller sets of similar pumps were erected by Messrs. Gwynne in the previous year. At the trials of these pumps they each discharged 88 tons of water per minute while going at 100 revolutions; the coal consumption being at the rate of 5·94 lb. per hour per horse-power of water lifted. Pumps erected by the same firm at Dordrecht, lifting 54 tons of water a minute 6·56 feet high, running at 120 revolutions a minute, took 24·3 horse-power, and consumed at the trial, made in 1876 and lasting over 1¾ hours, 6·69 lb. of coal per hour per horse-power of water lifted. Two smaller pumps, discharging together 65¼ tons a minute to the height of 6·56 feet, with 125 revolutions per minute, required 29·4 horse-power, and consumed at the rate of 7½ lb. of coal per hour per horse-power of water lifted.

WATERLAND, HOLLAND.—This polder is situated near Amsterdam, and contains 25,000 acres. There are three pumping stations provided with engines driving scoop wheels. The Buiksluit station has two curved scoop wheels, with the concave side of the curve towards the internal water. The curve is struck to a large radius, and, although

Pumping Stations. 153

this form is advantageous in entering the water, is not well adapted for parting with it. The wheels are 20·66 feet in diameter by 3·70 feet wide, and have sixteen curved iron scoops. The wheel makes $5\frac{1}{8}$ revolutions a minute to sixty of the engine, the speed at the periphery being 5·66 feet per second. The dip is about 4 feet, and the lift from 2·60 feet to 4·25 feet, the mean being 3·60; each wheel can discharge 105 tons a minute. The wheels are driven by two single-cylinder condensing horizontal engines of 30 actual horse-power. The cylinders are 1·37 feet in diameter, with 2·16 feet stroke. Steam is generated in two double-flued Lancashire boilers, each 6·87 feet by 28 feet long, having twelve Galloway tubes. The safety valve is weighted to blow off at 90 lb. The machinery was erected by the Prins Van Orange Company in 1875. The wheels work on an average 120 days of fifteen hours in the year.

One boiler is sufficient to generate steam for the two engines when discharging 210 cubic metres per minute. With the boiler first supplied, the consumption of coal was at the rate of 12·76 lb. per water horse-power. The ground in this polder is principally moor, and the water contains a great deal of sulphur and saltpetre, which had a very bad effect on the boilers, which were worked out in eight years. With the new boilers working at a pressure of 90 lb., the coal consumption has been reduced from about 4 to $2\frac{1}{4}$ tons per day of fifteen hours. The average consumption now is at the rate of 7·15 lb. per horse-power of water lifted.

MINDEN, HOLLAND.—The illustration given in Plate 7 is an example of a centrifugal pump for the drainage of land, driven directly by a vertical engine. The pump was made and sent to Holland by Mr. Hett, of Brigg, and is 36 inches in diameter, with pipes 10 inches in diameter. The speed of the engine is about 120 revolutions a minute. It is calculated to discharge 1700 gallons ($7\frac{1}{2}$ tons) a minute to a height of 6 feet. All the working parts are of forged steel, including the crank, which is counterbalanced and cut out of the solid. Full lubricating arrangements are provided, so that the engine can run for almost any length of time without stopping. The engine has a shifting eccentric for varying the cut-off. The cost of the pump, engine, and pipes was 250*l.* The engine and pump was erected and the boiler supplied by the Machinefabriek at Breda, and is reported by the proprietors as keeping their land dry in a very satisfactory manner.

KATATBEH AND ATFEH, IN EGYPT.—These pumping stations were formed for lifting the water from the Nile for irrigating the

154   *The Drainage of Fens and Low Lands.*

land and providing Alexandria with water. These two stations form the largest installation of pumping machinery in the world.

For the purpose of improving the irrigation of the Behera, or West District of the Nile, in Lower Egypt, a canal had been made communicating with the Nile by an intake above the Great Barrage, situated at the junction of the Rosetta and Damietta branches. For about six months in the year, or during low Nile, the river is not high enough for the water to flow into this canal; and it being excavated for the first 25 miles through desert sand, the supply was found to be inadequate for the irrigation of the lands and for the supply of Alexandria. In 1856 the Egyptian Government, in order to remedy the deficiency, decided to erect pumping stations at Katatbeh and Atfeh, so as to be able to pump into the canal from the Nile at a point below that portion of the canal which passes through the desert. The Katatbeh pumping station is situated about 25 miles north of the Barrage. The canal is 75 miles long, and discharges into the Mahmoudieh. The latter has its intake at Atfeh, 15 miles south of Rosetta, and supplies Alexandria and irrigates the north of the province. A large part of the traffic to and from Alexandria and the interior is carried on this canal.

In order to keep up the required supply it is necessary that the machinery at Atfeh should be worked during six months of the year, and that at Katatbeh four months.

Atfeh.—The machinery as originally constructed at this station consisted of four groups of single-cylinder beam engines, working centrifugal pumps by means of toothed wheels and pinions. These pumps raised 800,000 cubic metres in twenty-four hours ($555\frac{1}{2}$ tons per minute) to a height of 8 feet 6 inches. The results attained not being satisfactory, the machinery was altered from time to time by Messrs. Easton and Co., the principal change being the compounding of the engines. As, however, even after the alterations, the pumps could not discharge the quantity required (1369 tons a minute), they were finally removed and replaced in 1885 by eight scoop wheels. Four of the wheels were worked by the old engines: two by engines removed from Katatbeh; and the others by new compound engines.

Two of the wheels are 9·84 feet, and two 11·81 feet wide, with a diameter of 33 feet. Each wheel has eighty flat scoops 7·55 feet long, placed at an angle dropping from the radial line, being tangents to a circle concentric with the wheel 6 feet 6 inches in diameter.

The ordinary speed is 2·29 revolutions a minute, equal to a speed at the periphery of 4 feet per second. The other four wheels are the same diameter, 11·81 feet wide, with scoops 6 feet 7 inches long, and dip 4 feet below low-water. The speed is 1·91 revolutions a minute, equal to 3 feet a second. The engines make 12 revolutions to one of the wheel. The wheels work in a masonry raceway rendered with cement. The clearance between the wheel and the raceway is $\frac{3}{16}$ inch. The crest of the raceway is movable, and can be adjusted to suit the height of the water in the canal.

The eight wheels discharge at low-water of 0·30 metre, 2,922,708 cubic metres in twenty-four hours, equal to about 2000 tons per minute. The consumption of coal for the year 1885 is stated by Baghos Nubar Bey, in a paper in 'Trans.' Instit. C.E., to be 3·10 lb. per water horse-power per hour.

These wheels differ from those constructed in this country in having the scoops placed much closer together, their great width, and slow speed. The results attained appear to be very satisfactory, the consumption of coal as given being small as compared with that of the Dutch or English wheels.

Katatbeh.—The machinery originally erected for raising the water at this station was put up by Messrs. Easton and Co. in 1882. It consisted of a set of ten Archimedean screw pumps, placed parallel with each other, and worked by a shaft 164 feet long, coupled to three steam engines. These screw pumps were 39 feet long by 10 feet diameter, the cores being 4 feet diameter. The casing, core, and spiral strut were of wrought-iron plates. The spiral had a pitch of 25 feet. The screws made from 5 to 6 revolutions a minute, and discharged 25 cubic metres per revolution (about 136 tons per minute). These screws were found unequal to the enormous weight of water, and the cores broke at about one-third of their height. Seven of these were subsequently removed, the remaining three being kept as a reserve after being strengthened by tension rods placed radially, so as to connect the core and the casing. The weight was relieved by a ring of rollers fixed at two-thirds of the height, which turned on a roller path fastened to the masonry. The screws were replaced in 1883 by five vertical-action centrifugal pumps constructed by Messrs. Farcot and Co., each driven by a separate engine. The new engines are placed on the foundations prepared for the sluices of the screws, and the pumps in the basin below. The vertical shaft of the pumps has a crank to which the rod of the engine is directly connected. The lift varies from 20 inches to 10 feet, and the minimum duty of each pump was fixed by the contract at 500,000 cubic metres, raised to a maximum height of 3 metres in 23 hours, equal to about 2140 tons a minute 10 feet high. The whole, including the reserve screws, can discharge 3,500,000 cubic metres per day of 23 hours (about 2500 tons a minute), and the power in water raised is equal to 2000 water horse-power. The engines are of the Corliss type, having cylinders 39·37 inches in diameter, with 5 feet 10¾ inches stroke. The body of each pump is 19 feet 8 inches in diameter and 12 feet high. It stands on a group of six cast-iron columns, resting on the invert of the basin; regulating screws are fitted to each column to adjust the level. The discs or fans are 12 feet 5½ inches diameter by 4 feet 9 inches high. The average speed is 33 and the maximum 36 revolutions per minute. The water outlet, which springs from the annular chamber in the body of the pump, is formed of two cast-iron pipes 27 feet long. The inlet is trumpet-mouthed, 6 feet 10½ inches in diameter at the smallest part, enlarging to 9 feet 10 inches, with a further curved lip 10 feet 8 inches in diameter. Within the opening is an inverted cone, which, rising in the centre, passes from a diameter of 11·8 inches to 23·6 inches. The water attains a speed of 6 feet per second for the normal discharge of 6 cubic metres at 10 feet lift. The lower part of the pump shaft is made hollow, to admit a support solidly attached to the invert, and serving as a bearing above the level of the pump itself, and nearly 5 feet above the highest probable water-level. Above the disc the hollow shaft traverses a stuffing box in the dome of the pump. On the top of the iron column is fixed a cast-iron cylindrical head, 14·76 inches in diameter and 17·72 inches high. This head, on the upper face of which is a bearing of phosphor bronze, with a concave surface, supports the whole of the turning load, and also serves to centre the hollow shaft, the enlarged cavity of which is at the point provided with a bronze bush 10·82 inches deep. Above this bush the exterior shaft of cast iron divides into two branches, meeting again above, and leaving a wide opening for giving access to the pivot. The head of the hollow shaft above this opening is enclosed in a plummer block bolted to a heavy girder spanning the basin.

The quantity of coal to be consumed was limited by the terms of the contract to 3·85 lb. best English coal per water horse-power per hour, the quantity actually consumed being 3½ lb. The engines give off 65 per cent. of the indicated horse-power in water actually lifted.

A full description of these works, with drawings and illustrations, will be found in a paper on Irrigation in Egypt, in 'Le Génie Civil,' vol. x., 1886. An abstract of this paper will be also found in the 'Transactions' of the Institution of Civil Engineers, vol. lxxxix., 1887 ; and also in 'Engineering' of Jan. 28, 1887, where illustrations of the engine and pump at Katatbeh are given.

APPENDIX.

TABLE I.

EXPLANATION OF TERMS USED.

The quantities throughout are expressed, unless otherwise specially mentioned, as given below:—

The *mean velocity* of the current in feet per second.

Quantity in cubic feet or tons; a cubic foot of water being taken as 62·50 lb., and a ton 35·84 cubic feet.

Rate of discharge = cubic feet per second.

Hydraulic mean depth = the product found by dividing the sectional area of a channel by the perimeter or wetted contour, that is, length of the sides and bottom in contact with the water. This is sometimes termed the hydraulic radius.

Fall = the rate of inclination in the surface of the water in feet per mile.

Head = the difference in the level of the surface of the water at the upper and lower side of the place of the discharge in feet; *e.g.* in an engine drain the head or height which the pump has to lift the water is the vertical distance between the level of the surface of the water in the drain on the inside, and that of the channel on the outside, into which it is discharged. In a submerged sluice the head is in the same way the difference of level in the surface of the water on the inside and outside.

W.H.P. Horse-power in water actually lifted and discharged.

TABLE II.

WEIGHTS AND MEASURES RELATING TO WATER.

1 gallon of water weighs	10	lb.
1 ,, seawater ,,	10·20	,,
1 cubic foot of water ,,	62·32	,,
(generally taken as 62·50 lb.)		
1 ton of water .. .. .. =	35·84 cubic feet.	
do. do. .. .. .. =	224 gallons.	

Appendix.

1 cubic foot of water	=	0·0279017 ton.
do. do.	=	6·232 gallons.
1 gallon	=	·16051 cubic foot.
do.	=	277·274 „ inches.
1 cubic metre of water	=	·985 ton.
437·5 grains	=	1 oz. avdp.
7000 „	=	1 lb. „

French Weights and Measures, and their Equivalents in English Terms.

1 foot	=	0·30479545 metre.
1 metre	=	3·280899 feet.
1 inch	=	2·539954 centimetres.
1 centimetre	=	0·3937079 inch.
1 square foot	=	0·09289968 metre.
1 „ metre	=	10·7643 feet.
1 „ inch	=	6·45016, &c., sq. centimetres.
1 „ centimetre	=	0·154757, &c., sq. inch.
1 cubic foot	=	0·0283153 cubic metre.
1 „ metre (stere)	=	35·31658 „ feet.
1 cubic yard	=	0·764513 „ metre.
1 „ metre	=	1·30802 „ yards.
1 kilometre	=	0·6214 mile.
1 mile	=	1·60932 kilometres.
1 grain (avdp.)	=	0·064798, &c., gramme.
1 ounce (oz.)	=	28·349375 grammes.
1 pound (lb.)	=	0·45359265 kilogramme.
1 gramme	=	15·43267 grains.
453·59 grammes	=	0·352739 ounce.
1 kilogramme	=	2·20485 lb.
1 ton	=	1016·0475 kilogrammes.
1 gallon	=	4·543457 litres.
1 litre	=	0·2200967 gallons.
1 hectolitre	=	22·00967 gallons.
1 bushel	=	36·34766 litres.
1 hectolitre	=	2·7512 bushels.
1 cubic foot	=	28·33 litres.
1 perch of land	=	·2529194 are.
1 are	=	3·95383 perches.
1 acre	=	·404671 hectare.
1 hectare	=	2·471143 acres.

1 cubic metre of water   =   0·985 ton.
1 hectolitre of coal   .. = about 2¾ bushels, and weighs about 1½ cwt.
1 kilogramme per square centimetre   .. .. = 14·22128, &c. lb. per square inch.

TABLE III.
For ascertaining the pressure and velocity of water.

The pressure in lbs. per square foot .. .. = height in feet multiplied by 62·5 lb.

The pressure in lbs. per square inch .. .. = height in feet multiplied by 0·4335 lb.

Height in feet .. .. = pressure per square inch multiplied by 2·307.

The velocity of water theoretically is found by the formula $V = \sqrt{2gh}$, where V = velocity in feet per second; g, the velocity of a body falling freely for one second, or 32·182 feet. h = the height or vertical distance in feet through which the body passes. This is generally taken for practical purposes as eight times the square root of the head.

The theoretical velocity has to be reduced for loss by friction, &c., varying according to the character of the opening of channel through which the water passes. The result, as found above, therefore, has to be multiplied by one of the following constants which have been determined by experiment:—

Formulæ $V = 8\sqrt{h \times c}$, and

$$h = \left(\frac{V}{8 \times c}\right)^2.$$

h = height in feet. In the case of sluices or bridges, this is the vertical difference in the surface of the water above and below the opening.

V = velocity in feet per second.

c = constant.

  = opening of a bridge or sluice with pointed piers and smooth masonry   .. .. ·96

= opening of a bridge or sluice with square piers, or
where the doors are so hung as to cause eddies .. ·86
= small openings of sluices, as in a lock, whether submerged or not, or sluices with tankard-lid doors
causing resistance on all four sides .. .. .. ·64

These constants have been selected as appearing to give the most correct result.

TABLE IV.

To find the velocity of water in open channels.

$V = (\sqrt{R \times F_2}) C.$

V = velocity in feet per second.
R = hydraulic mean depth, or area of water divided by the length of the sides and bottom of the channel in direct contact with the water.
F = fall in feet per mile.
C = constant for friction, &c.
  = large streams .. .. ·90
  = large drains in good order .. .. ·80
  = secondary drains .. .. ·70
  = small drains .. .. ·60

The mean velocity of a stream is equal to the surface velocity in the centre multiplied by 0·83.

Example.—The mean velocity of a large drain having 10′·6″ bottom, 7′·6″ depth, slopes 2 to 1, and inclination 2 inches per mile:—

Area = 191·25 feet, contour 44·16 feet.

$$\frac{191·25}{44·16} = 4·33, \&c.$$

$V = (\sqrt{4·33 \times ·16 \times 2}) · 80$
$V = 0·9432$ feet per second.

TABLE V.

To find the cubic feet of water to be discharged by a drain or pumping engine per minute due to any given quantity of rain falling in 24 hours, and the horse-power in water raised for each foot of lift, multiply the number of acres drained by the multiplier opposite the rainfall.

Rain in 24 Hours.		Multiplier for Cubic Feet per Minute per Acre.	Multiplier for W.H.P. per Foot of Lift per Acre.
In parts of Inches.	Decimal Equivalents.		
$\frac{1}{8}$	·125	0·3151	·000597
$\frac{1}{4}$	·250	0·6303	·001194
$\frac{3}{8}$	·375	0·9453	·001790
$\frac{1}{2}$	·500	1·2606	·002387
$\frac{5}{8}$	·625	1·5758	·002985
$\frac{3}{4}$	·750	1·8909	·003581
$\frac{7}{8}$	·875	2·2060	·004178
1	··	2·5210	·004775

Example.—To find the quantity to be discharged per minute from an area of 10,000 acres with a rainfall of a quarter of an inch in 24 hours; and lift of 5 feet :—

10,000 × 0·6303 = 6303 cubic feet.

10,000 × 0·001194 × 5 = 59·70 water horse-power.

Rainfall.

1 inch rain = 3,630 cubic feet per acre.
„ „ = 101·183 tons „
„ „ = 122,614·90 gallons „

TABLE VI.

To find the discharge of water through sluices.

Multiply the area of the waterway, or the space occupied by the water, by the velocity as found in Table III.

In calculating the area of the water passing through the sluice, the depth must be taken at the lower side of the opening, as the velocity due to the head is not acquired until the lowest level is reached. In a tidal sluice, both the depth of water and velocity will vary from the time when the doors are first opened till they are closed again by the rising tide. The calculation must therefore be made on the mean velocity of the water.

Example.—A sluice with pointed piers, and doors fully open, allowing a free discharge of the water, having 16 feet opening, with depth of water 9 feet, and the head, or difference in the level of the surface above and below the sluice, of 2 inches, or say 0·16 foot, and allowing a deduction of 4 per cent. for friction, would discharge 442·36 cubic feet per second.

$$\text{Area} = 16' \; 0'' \times 9' \; 0'' = 144' \; 0''.$$
$$\text{Velocity} = 8\sqrt{·16} \times ·96 = 3·072 \text{ feet per second.}$$
$$\text{Discharge} = 144 \times 3·072 = 442·36 \text{ cubic feet per second.}$$

If the water approached the sluice with a velocity of 1·50 feet per second, allowance would require to be made for the head due to this, then :—

$$\text{Head due to this} = h = \left(\frac{1·50}{·96 \times 8}\right)^2 = 0·038.$$

Adding this to the head through the sluice, the head = ·16 + ·038 = ·198, which gives a velocity of 3·4176 feet per second, and discharge of 144 × 3·4176 = 492·134 cubic feet per second.

TABLE VII.

TIDAL SLUICES.

To find the quantity of water due to 24 *hours' rainfall if discharged in a limited time, owing to the sluice being closed by the tide.*

Multiply the given discharge per hour, minute, or second, by the multiplier opposite the number representing the hours the sluice doors are open each tide.

Example.—The discharge from the rainfall of 24 hours over a given area is equal to 500 cubic feet per minute. This has to pass through a sluice, the doors of which are closed 5 hours each tide, or 10 hours out of the 24, consequently the discharge will have to be effected in 14 hours, or 7 hours each tide, and the quantity per minute will be 855 cubic feet :—

$$500 \times 1·71 = 855.$$

No. of Hours Sluice Doors open each Tide.	Multiplier.
12	1
11	1·09
10	1·20
9	1·33
8	1·50
7	1·71
6	2·00
5	2·40
4	3·00
3	4·00
2	6·00
1	12·00

TABLE VIII.

To ascertain the time the doors of a sluice are closed by the tide.

Having ascertained the level of the sill of the sluice with reference to low-water in the sea or estuary into which the water is usually discharged, divide the height of the water in the drain by the range of the tide; the constant in the table nearest the result will give the time the doors will close with the rising tide and open again on the falling tide.

Time after High Water. Falling Tide. H.M.	Constant.	Time before High Water. Rising Tide. H.M.
0·0	1·00	6·00
0·30	·96	5·30
1·00	·92	5·00
1·30	·84	4·30
2·00	·75	4·00
2·30	·63	3·30
3·00	·50	3·00
3·30	·38	2·30
4·00	·26	2·00
4·30	·16	1·30
5·00	·08	1·00
5·30	·025	0·30

Example.—Supposing a tide rising 22′ 0″ above low-water spring tides, and the water to run in flood at 10 feet above the sill at low-water, accumulating during tide time to 14 feet—then 10 divided by 22, the rise of the tide, gives 0·45, which, being nearly the mean between 2 hours 30 minutes and 3 hours, the time given in the table the doors will close 2¾ hours after flood, leaving 3¼ hours to high-water. Again, 14 divided by 22, gives 0·63, which from the table gives the time for opening again as 2½ hours after high-water, making the time the doors would be shut as 5¾ hours each tide or 11½ in the day. The drain must therefore be calculated to discharge the rainfall due to 24 hours in 12½ hours.

The figures in the table are based on the assumption that the outfall is on or near the sea, or an estuary, and that the ebb and flow both last about 6 hours. If the sluice is situated some distance up a tidal river, these figures will require modification, and the time determined by observation of the tide at the particular locality where the sluice situated.

TABLE IX.

To ascertain the height of the tide at any hour during the ebb and flood.

Rule.—Multiply the range of the tide for the day by the constant given in the table opposite the time required, the result will give the height of the tide at that time.

Example.—Required the height of a tide, the high-water level of which rises 22 feet above low-water, at 2½ hours after high-water, or the same time after flood. The constant for a falling tide opposite 2·30 is 0·63, which, multiplied by 22, gives 13·86 feet as the height at 2 hours after high-water. For a rising tide the constant is 0·38, which, multiplied by 22, gives 8·36 feet as the height above low-water 2½ hours after flood.

Falling Tide. Time after High Water. H.M.	Rising Tide. Time after Flood. H.M.	Constant for 6 Hours. Ebb and Flood.
0·00	6·0	1·00
0·30	5·30	0·96
1·00	5·00	0·92
1·30	4·30	0·84
2·00	4·00	0·75
2·30	3·30	0·63
3·00	3·00	0·50
3·30	2·30	0·38
4·00	2·00	0·26
4·30	1·30	0·16
5·00	1·00	0·08
5·30	0·30	0·025

TABLE X.

To find the horse-power required for pumping-engines.

H.P. = 33,000 lb. (14·73 tons) lifted 1 foot high per minute.

N.H.P.—The nominal horse-power of an engine is a commercial term, used solely for the purpose of giving a general idea of the power of an engine.

I.H.P.—The indicated horse-power is the power of the engine calculated from the mean pressure of the steam in the cylinder throughout the whole of the stroke, as obtained by indicator diagrams.

W.H.P.—The horse-power in water actually raised and discharged.

Efficiency.—The efficiency of pumping machinery is expressed by the percentage of useful effect in water raised of the power given off by the engine, as found by the indicator diagram. Thus, if the indicated horse-power is represented by 100, and 40 per cent. of this is absorbed in working the engine and pump, the efficiency of the machinery is 60 per cent., or adopting the decimal notation and taking perfection at 100, the efficiency would be ·60. The difference depends on the efficiency of the machinery; the best engines and centrifugal pumps give off as much as 60 per cent. of the indicated horse-power. Although the trials do not record the proportion of this due to the engine as separate from the other part of the machinery, of the difference, approximately 10 per cent. may be given to the engine, and 30 per cent. to the pump. In ordinary practice it would not be safe to reckon on a higher efficiency than 60 per cent. for the pump, or 50 per cent. for engine and pump.

To find the net horse-power (W.H.P.) required to lift any given quantity of water a given height per minute :—

1. Multiply the number of cubic feet of water by 62·5, and by the height in feet which it has to be lifted, and divide the product by 33,000;
2. Or, multiply the number of cubic feet of water by the height to be lifted in feet, and divide by 528;
3. Or, multiply the number of cubic feet by ·0019 (·00189393, &c.), and by the height to be lifted in feet. This gives a slight excess.

Examples.—To find the horse-power required to lift 500 cubic feet of water 10 feet high per minute :—

$$(1) \quad \frac{500 \times 62 \cdot 5 \times 10}{33,000} = 9 \cdot 47 \text{ W.H.P.}$$

$$(2) \quad \frac{500 \times 10}{528} = 9 \cdot 47 \text{ W.H.P.}$$

$$(3) \quad 500 \times \cdot 0019 \times 10 = 9 \cdot 50 \text{ W.H.P.}$$

To find the indicated horse-power required, that is, the power to lift the water and work the machinery, allowing an efficiency of 55 per cent :—

$$\frac{9 \cdot 47 \times 100}{55} = 17 \cdot 22 \text{ I.H.P.}$$

TABLE XI.

Showing the dimensions and capacity of drains and sluices, and the area of land for which they are adapted.

Bottom Width.		Depth of Water.		Fall per Mile.	Constant for Friction, &c.	Slope of Sides.	Discharge per Second.	Area Drained.	Width of Sluice required.	
ft.	in.	ft.	in.	in.			cubic ft.	acres.	ft.	in.
1	0	1	0	6	·60	¾ to 1	0·74	78	1	0
1	6	1	3	6	·60	,,	1·48	140	1	0
2	0	1	6	6	·60	1 to 1	2·99	280	1	6
3	0	1	6	6	·60	,,	3·92	370	1	9
4	0	1	9	4	·60	,,	5·17	500	2	0
6	0	2	0	4	·65	,,	9·72	950	4	0
8	0	2	6	4	·65	,,	17·92	1,700	6	0
10	0	2	9	4	·65	1¼ to 1	27·36	2,600	9	0
12	0	3	0	4	·65	,,	37·09	3,530	10	6
14	0	3	6	4	·65	,,	54·26	5,168	14	0
16	0	4	0	4	·70	,,	81·73	7,780	18	0
18	0	4	6	4	·70	1½ to 1	114·15	10,900	24	0
22	0	5	0	3	·75	,,	150·23	14,300	25	6
26	0	5	0	3	·75	,,	173·11	16,500	29	6
28	0	5	6	3	·75	,,	215·32	20,500	35	0
30	0	6	0	3	·80	2 to 1	298·37	28,400	42	0
32	0	6	6	3	·80	,,	361·53	34,400	45	0
34	0	6	9	3	·80	,,	403·67	38,400	47	6
36	0	7	0	3	·80	,,	450·80	43,900	50	0
38	0	7	6	3	·80	,,	521·85	49,700	53	6
40	0	8	0	2	·80	,,	617·17	58,700	56	0

This table gives approximately the dimensions of drains and sluices required for conveying the water off fens or polders.

168 The Drainage of Fens and Low Lands.

The calculations are based on a quantity of water due to a quarter of an inch rain in 24 hours, and allowing for the drains and sluices to run the whole 24 hours.

The larger sluices are calculated to discharge at about the same velocity as the drains. In the smaller sluices a greater head has been allowed. If a greater head can be given, the width of the sluices can be reduced accordingly.

As the velocity varies as the square root of the fall, four times the fall per mile given in the table will double the discharge, or, one-fourth of the fall will give one-half the discharge.

INDEX.

A.

Adria, scoop wheel at, 87
Aeration of the soil by drainage, 38
Agricultural Show, trials of engines at, 61
Air spaces in drained soils, 37
Airy, Sir G. B., on scoop wheels, 72
—— W., on scoop wheels, 84
—— —— on screw pumps, 90
Anderson, W., on Wexford Harbour machinery, 133
Angles of egress and ingress for scoops, 75
Appleby & Co., machinery erected by, 125
Appold, J. G., centrifugal pump, 92, 118
Archimedean screw pump, 88
—— —— —— angle of tilt, 89
—— —— —— spiral angle, 90
—— —— —— discharging capacity, 89
—— —— —— comparative cost of, 53
—— —— —— efficiency, 90
—— —— —— at Katatbeh, 156
Ardizzoni, pump at Ferrara, 135
Arkelschendam, steam used for pump, 57
Atfeh, pumping station at, 155
—— quantity of water raised by wheels, 82
Atkinson, A., Butterwick engine, 127

B.

Ballyteigne Reclamation, sluices for, 28
Barker, on coal consumption, 62
Batter of drains, 15

Beemster Polder, pumps at, 152
Beijerinck, on scoop wheels, 75
Bijlermeer Polder, pump at, 150
Black Sluice drainage, 4
Boezem, or collecting basin, 141
Boilers, insurance of, 63
Boning rods, for setting out drains, 47
Bottisham Fen, steam used for draining, 58
Breasts, movable, for scoop wheels, 78
Brown, C., on pumps, 102
Bucket pumps, 50
—— —— comparative cost of, 53
—— —— at Lake Haarlem, 144
Bull Hassocks wheel, 130
Bullewijker Polder, pumping station at, 149
Burnt Fen, pumping station at, 119
Butterley Iron Company's engines, 58, 122
—— scoop wheels, 73
Butterwick pump, 127

C.

Carmichael, G., pumping engines, 120, 122
Catchwater drains, 7
Centrifugal pumps as compared with scoop wheels, 51
—— —— report on use of, by Dutch Commission, 51
—— —— comparative cost of, 53, 93
—— —— cost of, and engines, 67
—— —— description of, 92, 95
—— —— materials passing through, 94

Centrifugal pumps, points of a good pump, 96
—— —— angles of blades, 96
—— —— with horizontal spindles, 97
—— —— —— vertical spindles, 98
—— —— guide blades, 99
—— —— action of, 100
—— —— velocity of water through, 100
—— —— diagram of working, 101
—— —— efficiency of, 101
—— —— adaptability to varying heads, 93, 103
—— —— sizes and discharge, table of, 105
—— —— gearing for, 103
—— —— statical height of water supported by, 102
—— —— substances passing through, 94
—— —— description of, at Burnt Fen, 119
—— —— —— Bullewijker, 149
—— —— —— Bijlmermeer, 150
—— —— —— Beemster Polder, 152
—— —— —— Ferrara Marshes, 134
—— —— —— Fondi, 137
—— —— —— Fos, 135
—— —— —— Grootslag, 103
—— —— —— Gallejon, 137
—— —— —— Glassmoor, 126
—— —— —— Katatbeh, 156
—— —— —— Lade Bank, 110
—— —— —— Legmeer Polder, 104
—— —— —— Messingham, 127
—— —— —— Middel Polder, 104
—— —— —— Minden, 153
—— —— —— Prickwillow, 121
—— —— —— Redbourne, 129
—— —— —— Whittlesea Mere, 118
—— —— —— Wexford Harbour, 131
—— —— —— Zuidplas Polder, 150

Clearance of scoop wheels, 74
Coal used per h.-p. for engines, 61
—— quantity not proportional to lift, 62, 148
—— —— used depends on engine man, 61
—— —— by Dutch machinery, 61
—— Dutch contracts respecting quantity, 62
—— consumption of, at Halfweg, 62
Codigoro, centrifugal pump at, 94
Coode, Sir J., on rainfall, 9
Cooke, Mason, trials of Hundred Foot wheel, 114
Cost of pumping station, 66
—— for maintenance, 68
Cubitt, Sir W., on scoop wheels, 72
Cuppari, Sig., on water-raising machines in Holland, 52

D.

Danube river, deepening by scour, 22
Deeping Fen drainage engines, 106
—— —— deepening drain by scour, 20
De Witt, Messrs., drainage engines erected by, 54
Dirtness wheel, 130
Dorné, trials of pumps, 136
Drains, slope of sides, 15
—— depth and width of, 10
—— land occupied by, 16
—— proportion of width to depth, 10
—— example of size of, 25
—— table showing dimensions and capacity, 167
—— engine, 63
—— cleaning and removal of weeds, 17, 18
—— sizes of, 24
—— fall in surface, 14
—— —— field drain, 44
Drainage by gravitation, 3, 7
—— steam power, 50
—— field, 36
—— —— effect of, in drought, 37

Index.

Drainage, field, aeration of soil by, 38
—— —— temperature increased by, 38
—— —— increased value of land by, 41
—— —— time to put field drain in, 42
—— —— depth of, 43
—— —— distance apart, 44
—— —— direction and fall, 44
—— —— pipes, 48
—— —— cost of, 49
—— —— table of number of pipes required per acre, 49
—— —— setting out and levelling, 47
—— —— use of boning rods, 47
Dredging machinery for cleaning drains, 20
Drivers of engines, 60

E.

East Fen drainage, 3, 110
Easton & Anderson, as to cost of wheels and pumps, 54
—— —— exhibition of centrifugal pumps, 93
—— —— diagram for centrifugal pumps, 101
—— —— machinery erected by, 111, 119, 122, 126, 133, 155
Efficiency of scoop wheel, 85
—— centrifugal pump, 101
—— engines, 101
Egress, angle of, for scoop, 75
Engineer,' 'The, description of machinery, 24, 28, 102, 121
'Engineering,' description of machinery, 22, 157
Engines used for driving pumps and scoop wheels, 55
—— horse-power, description of meaning, 166
—— management of, 60
—— general description of, 60

Engines, attendants on, 61
—— simplicity of construction desirable, 58
—— power required for pumping, 64
—— tables showing W.H.P. required, 162
—— running of, at night, 64
—— coal consumed by, 61
—— drains, 63

F.

Fall in surface of drains, 11, 14
Farcot & Co., machinery erected by, 156
Fascines, training outfalls by, 31
Ferraby sluice, 30
Ferrara Marshes pumping station, 134
Field drainage, see Drainage
Float wheel, see Scoop Wheel
Floats, 74
Fondi pumping station, 137
Fos pumping station, 135
French weights and measures, 159

G.

Gainsborough, trials of portable engines at, 61
Gallejon pumping station, 137
Garonne river, 19
Gibbs & Deane, pumps for Lake Haarlem, 142
Glassmoor pumping station, 126
Glen river, 21
Glynn, J., 73
Gouda pumping station, 148
Gravitation, drainage by, 7
—— compared with steam power, 6
Grootslag Polder, pump for, 103
Guppy, T. R., description of Fondi station, 137
Gwynne & Co., machinery erected by, 151, 152

Gwynne, J. & H., description of centrifugal pump, 95
—— —— efficiency of their machines, 101
—— —— machinery erected by, 103, 104, 134, 135, 149, 150

H.

Haarlem, Lake, windmills, 57
—— description of drainage, 141
Haddenham Fen scoop wheel, 82
Halfweg pumping station, 147
—— coals used for different lifts, 62
Hamit, G., patent for scoop wheel, 82
Harrison, A., dredging machine, 23
—— —— Podehole engines, 109
Hatfield Chace scoop wheels, 129
Hathorn, Davey & Co., Burnt Fen pump, 121
Hawkshaw, Sir J., on rainfall, 8
—— —— Middle Level syphons, 34
—— —— Lade Bank pumps, 110
Head or lift of scoop wheel, 84
Heathcote, J. M., on pumps and scoop wheels, 54
Hett, L., centrifugal pumps, 102, 127, 129, 153
Holland, drainage in, 2
Horse-power of engines, 166
Huet, A., 87
Hull river, 13
Humber river, 19
Hundred Foot pumping station, 113
Hunt, Professor, on effect of drainage, 37
Hydraulic mean depth, 12, 25, 161

I.

Inclination of surface of water, 14
Ingress, angle of, for scoops, 75
Insurance of boilers, 63

K.

Katatbeh, description of machinery, 153
Katwijk pumping station, 148
—— quantity of water raised by wheels, 82
Kilmore Reclamation, sluices used at, 28
Kingston, J., scouring machine, 22
Korevaer, Mr., 87, 90

L.

Lade Bank pumping station, 94, 110
Ladles, 74
Land occupied by drains, 16
—— improvement of, by drainage, 41
—— table showing capacity of drains and sluices for draining, 167
Leeghwater, drainage of Lake Haarlem, 141
—— engine, Lake Haarlem, 144
Legmeer Polder, pump at, 104
Levelling rods for field drains, 47
Lift of scoop wheels, 84
Lifts, coal used for different, 62
Lincolnshire Agricultural Society, trials of engines, 61
Littleport and Downham drainage district, 112
—— —— windmills used in, 56

M.

Maas river, deepening by scour, 21
Machines used for dredging and scouring, 20
—— for raising water, 50
Madden, Dr., experiments on drainage, 40
Maintenance cost of drainage engines, 68

Index. 173

Management of drainage engines, 60
Mare Island Straits, deepening by scour, 24
Marozzo Marshes, pump for, 135
Marton drainage district, 50
Mersey river, deepening Pluckington Bank by scour, 22
Messingham pump and engine, 127
Middel Polder, pump at, 104
Middle Level drainage, 4
—— —— syphons, 34
Mijdrecht, steam engine used for drainage at, 57
Minden, pump at, 153
Molesworth's Pocket-book, 102
Morton Car scoop wheel, 79
Motion of water, 10
Movable breast for scoop wheels at Podehole, 79
—— —— Hundred Foot, 79
—— —— Katwijk, 80

N.

Natural drainage, 3
Naylor, S., patent for scoop wheels, 79
Night work for pumping engines, 64
North Level drainage, 4
North Sea Canal, pumps for, 112

P.

Paddles, 74
Parsons, W., on centrifugal pumps, 99, 100, 105
Perry, Captain, 56
Pipes for field drains, 49
Piston pumps, comparative cost of, 53
Po, valley of, 2
—— river, 19
Podehole pumping station, 106
Pompraden, or pump wheels, 149

Power of engine required for pumping, 64, 166
Pressure of water, 160
Prickwillow, pumping station, 121
Pumping stations, cost of, 66
—— —— cost of maintenance, 68
—— —— position of, 66
Pumps, bucket, 50

R.

Raceway for scoop wheels, 77
Rainfall, discharge of, 7
—— —— Lake Haarlem, 143
—— drainage due to, in main drains, 7
—— —— —— in field drains, 36
—— quantity of water to be pumped, 64
—— tables relating to, 162
Ravensfleet scoop wheel, 77
Redbourne pump engine, 129
Rennie, J., advised use of steam for draining fens, 58
—— Sir J., Ferraby sluice, 30
Rhine, river, 19
Rhone, river, 19
Richards, pumps used on Pacific coast, 101
Roding drains, 17

S.

Scoops, 73, 74
Scoop wheels, advantage of, as compared with other pumps, 51
—— —— report on, by Dutch Commission, 51
—— —— Cuppari's opinion of, 52
—— —— comparative cost of, 53
—— —— cost of engines, 67
—— —— first use of, 70
—— —— description of, 73
—— —— with curved scoops, 76
—— —— diameter, 81
—— —— angle of egress and ingress, 75

Scoop wheels, inlet and outlet courses, 77
—— —— shuttle for, 80
—— —— movable breasts, 78
—— —— quantity of water raised by, 82
—— —— speed of, 83
—— —— work done by, 84
—— —— efficiency of, 85, 115
—— —— clearance, 74
—— —— discharge of, 84
—— —— description of, at Podehole, 106
—— —— —— Littleport and Downham district, 112
—— —— —— Ten Mile station, 117
—— —— —— Hundred Foot, 113
—— —— —— Upwell and Outwell, 124
—— —— —— Burnt Fen, 119
—— —— —— Prickwillow, 121
—— —— —— Hatfield Chace, 130
—— —— —— Katwijk, 148
—— —— —— Halfweg, 147
—— —— —— Zuidplas, 150
—— —— —— Waterland, 152
—— —— —— Atfeh, 155
Scour, removal of deposit by, 18
Screw pumps, 88
Shuttle for scoop wheels, 80
Sluices, size of, 26, 167
—— doors, 27
—— for lateral drains, 28
—— foundations of, 29
—— at Ferraby, 30
—— position of, 31
—— level of sill, 31
—— capacity for discharge, 32
—— formula for calculating discharge, 33, 163
—— proportion of depth to width, 33
—— Stoney's improved method of hanging doors, 28
—— examples of, 30, 34

Sluices, table showing dimensions and capacity, 167
—— —— —— quantity discharged by tidal, 163
—— —— —— time closed by tide, 164
Soc, 36
South Level, drainage of, 5
—— —— cost of pumping in, 68
Spalding, trials of portable engines at, 61
Speed of scoop wheels, 83
Start posts, 73
Steam engines for drainage, 58
—— power, drainage by, 50
—— —— compared with gravitation, 6
—— first applied to drainage, 57
Sterk, Elink, trials of pumping machinery, 147, 149
Stoney, improvement in sluice doors, 28
Stour, river, deepening by scour, 21
Sturton-on-Trent scoop wheel, 77
Suspension, material carried in by water, 19
Syphons used at Middle Level drain, 34

T.

Tees, river, 19
Temperature of drained land, 38
Ten Mile pumping station, 117
Thompson, on centrifugal pumps, 102, 105
Tidal outfalls, 9
Tide, tables showing time sluices closed by, 164
—— —— —— height of, 165
Training outfalls by fascines, 31
Transporting power of water, 18
Trials of engine drivers, 61

U.

Unwin, centrifugal pumps, 102, 105
Upwell, Outwell, &c., wheel, 124

V.

Velocity of water in open drain, 11
—— mean, 13
—— formula for ascertaining, 12, 160
—— scoop wheels, 83
Vernatts drain, deepening by scour, 20
Vistula, river, 19
Vreedenberg, A. C. J., trials of pumps, 136

W.

Waldersea drainage district, 50
Water, quantity to be raised by pumping engines, 64
—— —— table, 162
—— —— raised by scoop wheels, 82
—— —— due to rainfall, table, 162
—— transporting power of, 18
—— sediment carried in suspension, 19
—— motion of, 10
—— velocity of, in open channels, 11, 161
—— —— through bridges and sluices, 32, 160, 162
—— pressure of, 160
—— inclination in surface, 14
—— machines used for raising, 50
Waterland, pumping station at, 152
Watt & Co., machinery erected by, 109, 114, 130
Watt & Co., trials of, as to discharge of wheels, 115
Weeds, obstruction from, 13
—— —— in river Hull, 13
—— removal of, 17
Weights and measures, 158
—— —— French, 159
Welland, river, deepening by scour, 21
Wells, use of Appold pump, 93
Welsh, E., trials of Lade Bank engines, 112
Wexford Harbour, cost of running pumps, 54
—— —— drainage stations, 131
Wheel pumps, 74, 149
Wheler, W., scoop wheel, 70
Whittlesea Mere, use of centrifugal pump, 93
—— pumping station, 118
Windmills, use of, for drainage, 55, 56, 57
—— for draining Lake Haarlem, 141
—— Zuidplas Polder, 150
Witham drainage, 4
—— river, 19

Y.

Young, Arthur, description of windmill in fens, 56

Z.

Zuidplas Polder pumping station, 57, 150

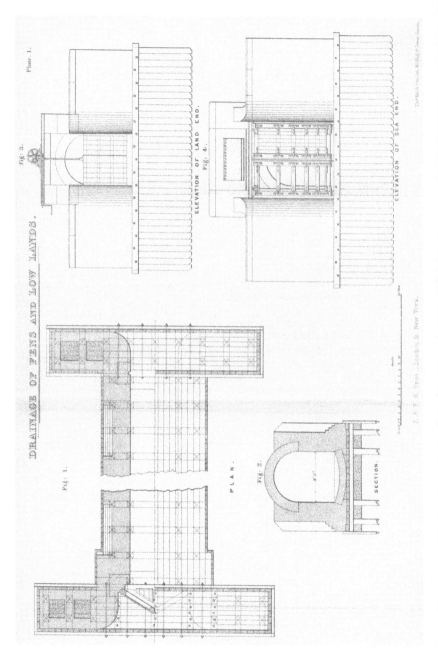

The material originally positioned here is too large for reproduction in this reissue. A PDF can be downloaded from the web address given on page iv of this book, by clicking on 'Resources Available'.

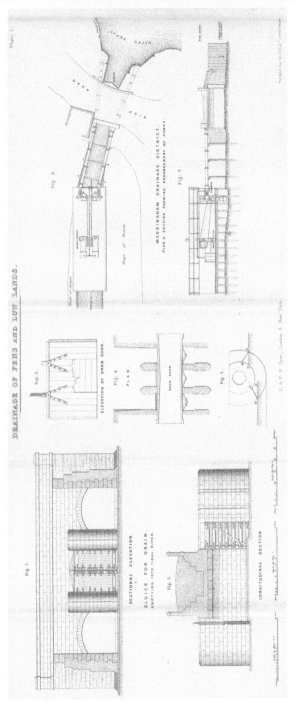

The material originally positioned here is too large for reproduction in this reissue. A PDF can be downloaded from the web address given on page iv of this book, by clicking on 'Resources Available'.

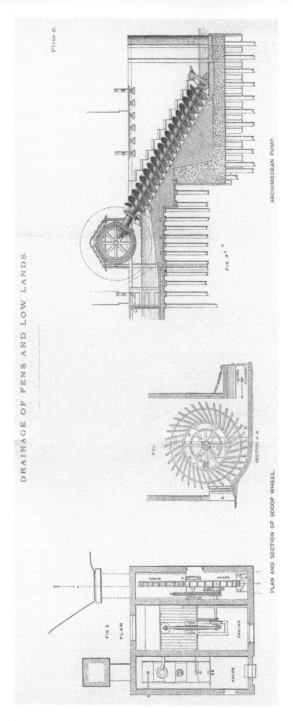

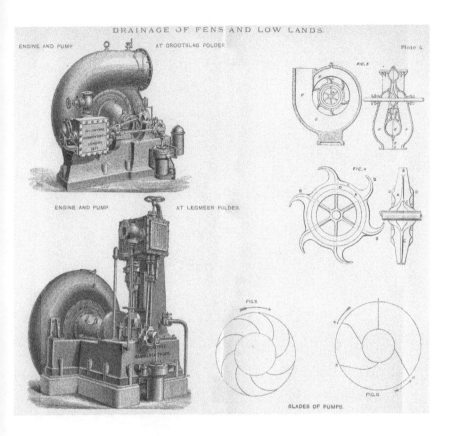

The material originally positioned here is too large for reproduction in this reissue. A PDF can be downloaded from the web address given on page iv of this book, by clicking on 'Resources Available'.

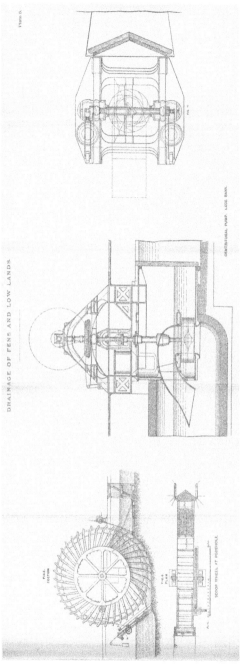

The material originally positioned here is too large for reproduction in this reissue. A PDF can be downloaded from the web address given on page iv of this book, by clicking on 'Resources Available'.

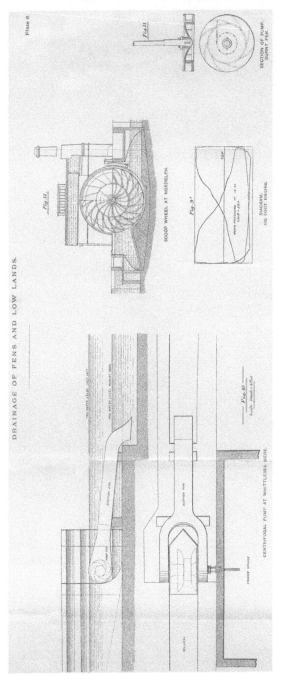

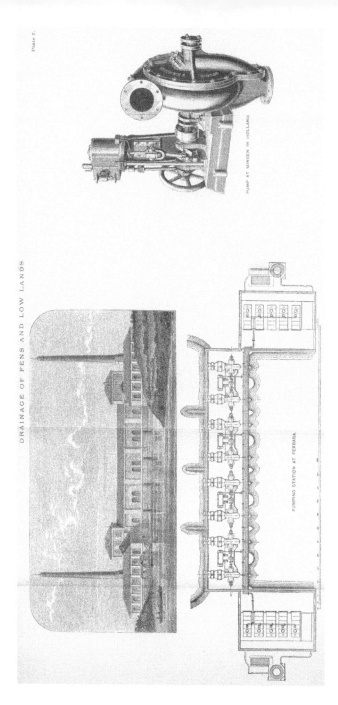

The material originally positioned here is too large for reproduction in this reissue. A PDF can be downloaded from the web address given on page iv of this book, by clicking on 'Resources Available'.

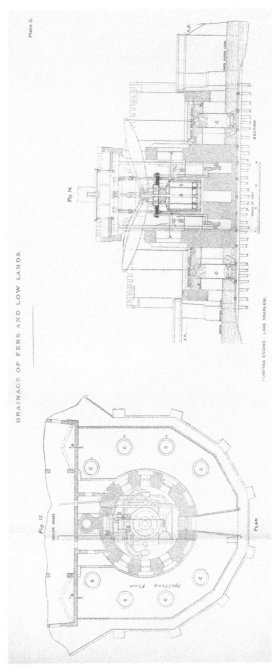

The material originally positioned here is too large for reproduction in this reissue. A PDF can be downloaded from the web address given on page iv of this book, by clicking on 'Resources Available'.

For EU product safety concerns, contact us at Calle de José Abascal, 56–1°, 28003 Madrid, Spain or eugpsr@cambridge.org.

www.ingramcontent.com/pod-product-compliance
Ingram Content Group UK Ltd.
Pitfield, Milton Keynes, MK11 3LW, UK
UKHW012344130625
459647UK00009B/515